Psychopharmacology for Medical Students

Arash Ansari, M.D. and David N. Osser, M.D.

authorHOUSE®

AuthorHouse™
1663 Liberty Drive
Bloomington, IN 47403
www.authorhouse.com
Phone: 1-800-839-8640

First published by AuthorHouse 8/11/2009

ISBN: 978-1-4389-9885-5 (e)
ISBN: 978-1-4389-9883-1 (sc)

Printed in the United States of America
Bloomington, Indiana

This book is printed on acid-free paper.

Dedication:

To our parents, wives, and children—
for their many sacrifices.

AA
DNO

TABLE OF CONTENTS

LIST OF TABLES

IMPORTANT NOTE: The information presented in this manuscript is meant to be an overview of major topics in psychopharmacology. This book is meant to be an introduction to the field, not a handbook for the administration of available psychotropics. Specifics regarding clinical use of medications including doses are presented for purposes of teaching and learning only. Although every effort has been made to present the material accurately, we cannot rule out typographical or other errors. As always, the package insert of each medication should be reviewed prior to administration, and treatment should be customized to the needs and characteristics of the individual patient after a thorough psychiatric evaluation.

INTRODUCTION

The use of psychotropic medicines to treat psychiatric illness has increased dramatically in recent times. Although the biological etiologies of most psychiatric disorders are still unclear, effective pharmacological treatments have been developed over the past 50 years that have become part of the standard of care in the treatment of most major psychiatric disorders.

Psychiatric medications are part of the armamentarium of most practicing physicians, regardless of medical specialty. In the United States, although most severe types of mental illness are likely to be treated by psychiatrists, most prescriptions for psychotropics (e.g. anxiolytics and newer antidepressants) are written by non-psychiatrists. (Stagnitti 2008) Psychiatric medications are consistently prominent in the list of the top 200 most commonly prescribed medications, and in the top 20 pharmaceuticals in terms of sales in the United States. From 2003-2007 antidepressants, as a class, topped all other therapeutic classes for the overall number of dispensed prescriptions in the U.S.(IMS Health 2007)

As in the treatment of all medical disorders, a thorough evaluation must precede psychiatric diagnosis and subsequent psychopharmacological treatment. A complete history should be obtained and the patient should be examined. Medical or neurological etiologies that may contribute to the presentation of psychiatric illness should be identified and addressed. Nearly 10% of patients presenting with a psychiatric complaint will turn out to have a medical problem as the primary cause. (Hall, Popkin, et al. 1978) Active substance abuse, if present, should be treated before or at the same time that pharmacological therapies are initiated.

The clinician should then decide if the condition requires medication treatment. Mild to moderate anxiety and depression generally respond equally well to supportive interventions or psychotherapy.(APA 2004; Barkham and Hardy 2001; Cuijpers, van Straten, et al. 2009; King, Sibbald, et al. 2000) On the other hand, if the psychiatric disorder or symptoms are severe, or if psychosis, mania, or dangerousness are present, then psychopharmacological treatments (and referral to a psychiatrist) are indicated. Although many primary care physicians may be quite comfortable with their ability to manage psychiatric illness, the amount of monitoring that is required to provide adequate follow-up should be taken into account before initiating treatment. When treating moderate to severe psychiatric illness, optimum therapy includes the use of concomitant psychotherapy in addition to pharmacotherapeutic measures.(APA 2004; APA 1998; APA 2000; Keller, McCullough, et al. 2000; Banerjee, Shamash, et al. 1996; Reynolds, Frank, et al. 1999; Katon, Von Korff, et al. 1999; Miklowitz 2008)

Placebo-controlled randomized clinical trials, using strict exclusionary criteria when selecting subjects, have traditionally been used to study a psychiatric medication's *efficacy* (i.e. the ability of the medication to treat the condition better than placebo under controlled conditions). For example, studies comparing an antidepressant to placebo may use an 8 week double-blind parallel design and include subjects with major depression but without any other medical or psychiatric co-morbidities. Response may be defined as a 50% improvement in a chosen outcome rating scale. These efficacy studies also provide the response data that pharmaceutical companies must submit to the Food and Drug Administration (FDA) to obtain indications for developed drugs.

Effectiveness studies, on the other hand, are often larger, naturalistic studies that attempt to approximate 'real world' conditions by studying patients who may have psychiatric and medical co-morbidities, and by relying on broader outcome measures for assessing response. These studies may compare outcomes of treatment with multiple medications. As such, effectiveness studies complement our understanding of drug efficacy.(Summerfelt and Meltzer 1998) Recent National Institute of Mental Health (NIMH) sponsored effectiveness studies have the added benefit of funding from a neutral (non-pharmaceutical industry) source, thereby avoiding possible study design shortcomings or evaluator biases that may influence study results.(Heres, Davis, et al. 2006; Osser 2008) These studies include (1) the Clinical Antipsychotic Trials of Intervention Effectiveness (CATIE),(Keefe, Bilder, et al. 2007; Lieberman, Stroup, et al. 2005)

(2) the Sequenced Treatment Alternatives to Relieve Depression Study (STAR*D),(Rush, Trivedi, et al. 2006; McGrath, Stewart, et al. 2006; Nierenberg, Fava, et al. 2006; Trivedi, Fava, et al. 2006; Fava, Rush, et al. 2006) (3) the Systematic Treatment Enhancement Program for Bipolar Disorder (STEP-BD),(Sachs, Nierenberg, et al. 2007; Goldberg, Perlis, et al. 2007; Miklowitz, Otto, et al. 2007) (4) the Clinical Antipsychotic Trials of Intervention Effectiveness—Alzheimer's Disease (CATIE-AD),(Schneider, Tariot, et al. 2006; Sultzer, Davis, et al. 2008) and (5) the National Institute of Alcohol Abuse and Alcoholism (NIAAA) sponsored Combined Pharmacotherapies and Behavioral Interventions Study (COMBINE).(Anton, O'Malley, et al. 2006; Anton, Oroszi, et al. 2008) Findings from these studies are now influencing clinical psychiatric practice.

In clinical practice, even after an appropriate diagnosis is made for an individual patient and the decision is made to use a medication from a particular pharmacotherapeutic class (for example an antidepressant for depression), multiple variables need to be considered prior to selecting a specific agent. The physician should take the following into account: (1) patient acuity and the need to address the most dangerous presenting symptoms (e.g. behavioral agitation, suicidality, catatonia, etc.) first, (2) the patient's past treatment history, (3) pre-existing medical conditions in order to minimize any increase in medical risk, (4) possible medication interactions, (5) the time required for amelioration of symptoms, (6) a medication's known side effect profile and how this may affect presenting symptoms, (7) the need to minimize the use of polytherapy, (8) possible pharmacogenetic factors and hereditary patterns of drug response and tolerance, and (9) financial cost-benefit considerations. The practicing physician should consider these issues prior to initiating treatment.(Ansari, Osser, et al. 2009)

Characteristics of the major classes of psychotropics and their use in adults are discussed below. Children and adolescents may tolerate or respond to these medications differently. The use of psychopharmacological therapies in these age groups is outside the scope of this book.

ANTIDEPRESSANTS

Currently available antidepressants primarily affect the norepinephrine and serotonin (monoamine) neurotransmitter systems.(Nestler, Hyman, et al. 2009) Norepinephrine systems originate primarily from the locus ceruleus (and lateral tegmental areas) and project widely to almost all areas of the brain and spinal cord. Serotonergic neurons reside in the raphe nuclei in the brainstem and diffusely make contact with all areas of the brain. Most antidepressants increase the available amount of norepinephrine and/ or serotonin at the neuronal synapse by decreasing the reuptake of these neurotransmitters into the pre-synaptic cell. They do this by inhibiting the norepinephrine transporter and/or the serotonin transporter, or by decreasing the metabolism of these neurotransmitters. Other antidepressants have direct effects on monoamine receptors. Genetic polymorphisms of the norepinephrine and serotonin reuptake transporters(Kim, Lim, et al. 2006) as well as polymorphisms of post-synaptic serotonin receptors(McMahon, Buervenich, et al. 2006) have been associated with differences in responses to different antidepressants. Once synaptic changes have taken place with treatment, long-term adaptations in post-synaptic neurons and resultant changes in gene expression may then be responsible for alleviating depression.(Nestler, Hyman, et al. 2009)

Tricyclic Antidepressants (TCAs)

Beginning with the introduction of imipramine in the late 1950's,(Kuhn 1958) tricyclic antidepressants were among the first classes of antidepressants developed. They share a tricyclic structure (two benzene rings on either side of a seven-member ring), exhibit variable degrees of norepinephrine and serotonin reuptake inhibition, and are antagonists at several other neurotransmitter receptors.(Hyman, Arana, et al. 1995) Examples of commonly used TCAs include the tertiary amines **imipramine**, **amitriptyline**, **clomipramine**, and **doxepin**, and the secondary amines **desipramine** (metabolite of imipramine) and **nortriptyline** (metabolite of amitriptyline). All TCAs can cause the following adverse effects: (1) slowing of intra-cardiac conduction as measured by QRS and QTc prolongation, (2) anticholinergic effects such as dry mouth, urinary retention, and constipation due to muscarinic acetylcholine receptor antagonism, (3) orthostatic hypotension due to peripheral alpha-1-adrenergic antagonism, and (4) sedation and possible weight gain due to histamine (H1) receptor antagonism. For these reasons TCAs need to be started at low doses and increased gradually, giving the patient time to accommodate to these effects. Individual differences in both severity of side effects and therapeutic effects (along with differences in therapeutic serum levels) (Perry, Zeilmann, et al. 1994) exist among individual TCAs. There is some evidence to suggest that TCAs may have a particularly important role in the treatment of depression with psychotic features. (Hamoda and Osser 2008)

The cardiac effects of TCAs have contributed to a reduction in their use over the past 20 years. Prolonged QT interval (measured as QTc when corrected for heart rate) may be associated with torsades de pointes, a potentially fatal ventricular arrhythmia (also see section on antipsychotics). All patients should have an ECG to rule out any existing conduction abnormalities prior to considering TCAs. Patients with recent myocardial infarctions should not initiate treatment with these antidepressants. Most importantly, depressed patients who are at risk for suicide and overdose may not be appropriate for treatment with TCAs. It should be noted that a 1-2 week supply of these medications can be fatal in overdose due to the risk of cardiac arrhythmias. Therefore, depending on the patient's risk of suicide, clinicians may need to limit the number of tablets prescribed with each refill. This concern is significantly lessened with the use of newer antidepressants that are safer in overdose.

TCAs are often used for their mild to moderate analgesic effects in the treatment of chronic pain syndromes.(Magni 1991) These effects are independent of any effect on mood, with efficacy starting at lower doses, and with response seen earlier than when antidepressants are used for depression.(Magni 1991; Max, Culnane, et al. 1987; Onghena and Van Houdenhove 1992) The TCAs seem more effective for chronic pain than the selective serotonin reuptake inhibitor antidepressants (SSRIs, see below).(Ansari 2000; Fishbain 2000; Saarto and Wiffen 2007)

Monoamine Oxidase Inhibitors (MAOIs)

Monoamine oxidase is an enzyme that acts to metabolize monoamines, both intracellularly and extracellularly. Its inhibition increases the amount of serotonin, norepinephrine and dopamine available for neurotransmission. The first MAOI, iproniazide, an anti-tuberculosis drug, was discovered in the 1950's. **Tranylcypromine**, **phenelzine**, **isocarboxazid**, and more recently **transdermal selegiline** are MAOIs currently available in the United States for the treatment of depression. These antidepressants may be particularly effective for patients with atypical depression (i.e. depression characterized by hyperphagia and hypersomnia).(Quitkin, Stewart, et al. 1993)

Although serotonergic side effects—see SSRIs below—as well as orthostatic hypotension can occur with MAOIs, there are two other primary areas of concern that limit the use of these agents.(Lippman and Nash 1990) First, dangerous interactions can occur with certain foods, such as aged cheeses and wines that contain biogenic amines such as tyramine. MAOIs can inhibit the metabolism of tyramine in the intestine, increasing its general circulation and ultimately leading to an increase in sympathetic outflow and an adrenergic ('hypertensive') crisis characterized by severe hypertension, headache, and increased risk of stroke and cerebral hemorrhage. Patients need to be advised regarding dietary restrictions before treatment. Also, to prevent hypertensive crisis, MAOIs cannot be combined with medications that have sympathomimetic properties such as some over the counter cold preparations, amphetamines, and epinephrine (which is often added to local anesthetics as a vasoconstrictor).

Secondly, MAOIs if used concomitantly with serotonergic agents, such as SSRIs, may lead to 'serotonin syndrome,' a potentially fatal condition that is characterized by hyperreflexia, hyperthermia, and tachycardia, and may lead to delirium, seizures, coma and death.(Sternbach 1991) A two week washout period is required when switching from SSRIs (or any other agents with serotonergic effects) to MAOIs, or vice versa. An exception is when the long half-life SSRI fluoxetine is being discontinued: a five week washout period is needed before starting an MAOI.(Boyer and Shannon 2005) The treating physician may access appropriate online databases such as the drug-specific DRUG-REAX® System (www.micromedex.com/products/drugreax) or GeneMedRx (www.genemedrx.com) to help assess potentially dangerous drug-drug interactions when considering the use of an MAOI.

Selective Serotonin Reuptake Inhibitors (SSRIs)

SSRIs are antidepressants with a more favorable side effect profile than TCAs and MAOIs and as such are used as first-line antidepressants. As their name implies, SSRIs inhibit the serotonin transporter from reuptaking serotonin at the neuronal synapse. Interestingly, polymorphisms at the promoter region of the serotonin transporter gene (SLC6A4) may influence response to SSRIs: the presence of the 'short' form of the serotonin transporter gene may be associated with poor response to SSRIs, whereas the presence of the 'long' allele may be associated with positive drug response(Malhotra, Murphy, et al. 2004) and better tolerability.(Murphy, Hollander, et al. 2004) Recent data from the large NIMH STAR*D study, however, have failed to support the association between this polymorphism and drug response.(Kraft, Peters, et al. 2007; Lekman, Paddock, et al. 2008)

Currently available SSRIs include **fluoxetine**, **paroxetine**, **sertraline**, **fluvoxamine**, **citalopram** and its S-enantiomer **escitalopram**. Possible mild early side effects (that can be minimized by starting the SSRI at a low dose and increasing the dose gradually) include gastro-intestinal upset, sweating, headaches, jitteriness or sedation. Continuation of these agents may be associated with reversible sexual side effects (i.e. delayed ejaculation, decreased libido, or erectile dysfunction) in 2-73% of treated patients (depending on how questions regarding sexual side effects are asked).(Montejo, Llorca, et al. 2001)

SSRIs differ in their propensities to inhibit hepatic cytochrome P450 enzymes (e.g. CYP1A2, CYP2C9, CYP2C19, CYP2D6, CYP3A4).(Ereshefsky, Jhee, et al. 2005) Inhibition of hepatic enzymes may lead to decreased metabolism of substrate medications such as warfarin, metoprolol, tricyclic antidepressants, and antipsychotics. This may increase serum levels of these drugs and lead to increased risk of dangerous adverse effects such as bleeding, hypotension, cardiac arrhythmias, and parkinsonian effects, respectively. Among the SSRIs, citalopram and escitalopram are the least likely to inhibit the metabolism of other drugs and are therefore preferred in patients concomitantly treated with multiple other medications. Escitalopram is a more recently marketed, and still expensive, successor to generic citalopram, though it may have no clinically significant efficacy advantage.

The relatively benign side effect profiles of SSRIs and their ease of use have contributed to widespread use by clinicians who might not have been comfortable with using earlier antidepressants such as TCAs and MAOIs. In cases of atypical presentations of depression, or depression in the context of recent substance abuse, SSRIs are more readily used even before there is absolute clarity in diagnosis. Under these circumstances, many clinicians believe that the benefits of treatment may outweigh the risks. In the case of depressed patients with concomitant substance abuse, the possibility that some patients may not be able to maintain sobriety *because* of an underlying major depressive disorder has served as a rationale for beginning antidepressant therapy even when the patient is still actively using.

Although this empirical 'trial' of an SSRI (as an antidepressant with a relatively benign side effect profile), in situations were there is less than optimum diagnostic clarity, may be appropriate for some patients, the physician should be aware of at least two major areas of risk. First, all antidepressants can induce mania in the short-term, and overall mood instability in the long-term, in patients with a vulnerability to bipolar disorder. A clear family history should be obtained to investigate whether there is a genetic predisposition to bipolar disorder. Also, clinicians should be aware that younger depressed patients, who may go on later to exhibit manic symptoms, may be incorrectly diagnosed with unipolar depression when in fact they may have a bipolar diathesis. A pre-bipolar presentation of depression(O'Donovan, Garnham, et al. 2008) should be suspected in patients with (1) a family history of bipolar depression, (2) a younger age of onset, (3) a family history of completed suicide, (4) past poor response to antidepressants, (5) a history of treatment-emergent agitation, irritability, or suicidality, and (6) a history of post-partum psychosis.(Chaudron and Pies 2003) Depressed patients with these characteristics may have bipolar rather than unipolar depression and therefore should not be reflexively started on an antidepressant.(Ghaemi, Ko, et al. 2002; O'Donovan, Garnham, et al. 2008; Phelps 2008)

Secondly, antidepressants have been associated with an increased risk of treatment-emergent suicidality—this occurs in about 4% of treated patients versus 2% on placebo—especially in children, adolescents and young adults as noted in the current package inserts of all antidepressants. It is still unclear if this risk is significant in adults over age 25. The reasons for this increase in suicidality are not clear, although increased agitation (e.g. akathisia) or activation as a side effect,(Harada, Sakamoto, et al. 2008) or the possible emergence of "mixed" manic symptoms (mania combined with dysphoric mood) in depressed bipolar patients as noted above, may be responsible. Despite the concern that antidepressants may infrequently increase suicide risk, it should be noted that overall rates of suicide in the United States had actually decreased over a prior 15 year span probably due to the increasingly widespread use of SSRI antidepressants. (Grunebaum, Ellis, et al. 2004) Nevertheless, the concern about treatment-emergent suicidality argues for the need for careful evaluation and diagnosis, increased discussion of risks and benefits of treatment with patients (and family when appropriate), and close monitoring of all patients beginning antidepressant therapy. Prescribing antidepressants when indicated, coupled with these steps, is more appropriate than withholding antidepressants in unipolar depressed patients who are more likely to benefit rather than come to harm from these treatments.(Bridge, Iyengar, et al. 2007) Unfortunately, recent surveys have found that instead of the increase that was hoped for in the monitoring of patients undergoing antidepressant therapy,(Morrato, Libby, et al. 2008) there has been an overall decrease in the use of antidepressants and a recent increase in the overall rates of suicide(Gibbons, Brown, et al. 2007) since the 'black box' warnings about treatment-emergent suicidality were issued.

Serotonin-Norepinephrine Reuptake Inhibitors (SNRIs)

The SNRIs, **venlafaxine, desvenlafaxine** and **duloxetine**, are dual action serotonergic and noradrenergic antidepressants that would be expected to have efficacy similar to TCAs though without anticholinergic, antihistaminic, hypotensive, or cardiac side effects. Venlafaxine is primarily serotonergic at lower doses and has a dual action only at higher doses. (Feighner 1999; Richelson 2003) Using venlafaxine at lower doses (i.e. less than 150 mg per day), therefore, should not be presumed to be any different from using an SSRI. At higher doses it can have a mild to moderate hypertensive effect,(Johnson, Whyte, et al. 2006; Mbaya, Alam, et al. 2007) although patients with effectively treated hypertension can tolerate venlafaxine without an increase in blood pressure(Feighner 1995). Duloxetine, which exerts a dual action effect throughout its dose range (i.e. not only at higher doses as with venlafaxine),(Stahl and Grady 2003) can also increase blood pressure, although the effect may be less pronounced and clinically insignificant.(Raskin, Goldstein, et al. 2003; Wohlreich, Mallinckrodt, et al. 2007) SNRIs, like TCAs, are more likely to induce mania in bipolar patients than SSRIs.(Leverich, Altshuler, et al. 2006)

Because low-dose TCAs have been shown to be modestly effective in the treatment of chronic pain syndromes, and SNRIs have a similar dual action, they have been proposed for the treatment of chronic pain symptoms as well. Duloxetine is currently the only antidepressant with an FDA indication for diabetic neuropathy and fibromyalgia. Although the more benign side effect profile of duloxetine may make it the preferred agent in a patient for whom the risks associated with a TCA are unacceptable, there is no evidence to suggest it would be more efficacious for the treatment of pain than the more cost-effective TCAs.

Antidepressants with Other Mechanisms

Bupropion is an antidepressant with a poorly understood mechanism of action. It is believed to exert its effect through dopamine reuptake inhibition although it is unclear why this mechanism alone should provide it with an antidepressant effect. Some data suggest that it may also exhibit norepinephrine reuptake inhibition.(Richelson 2003; Rosenbaum, Arana, et al. 2005) Bupropion has a different side effect profile than antidepressants that significantly affect the serotonergic systems. It is unlikely to cause sexual side effects or weight gain—two of the most common reasons for medication non-adherence in patients. However, bupropion can lower seizure threshold and is therefore contraindicated in patients who are seizure-prone (e.g. patients with a history of seizures or conditions that increase seizure risk, such as eating disorders, or in active withdrawal from alcohol or benzodiazepines). The risk of seizure is dose dependent: this should be kept in mind when combining bupropion with CYP2D6 inhibitors such as paroxetine or fluoxetine that may increase bupropion serum levels. Among antidepressants, bupropion is least likely to cause mania in bipolar patients.(Leverich, Altshuler, et al. 2006; Post, Altshuler, et al. 2006)

Mirtazapine increases both serotonin and norepinephrine at the neuronal synapse (and therefore like SNRIs has 'dual actions') through mechanisms distinct from reuptake inhibition. It is an antagonist at alpha-2-adrenergic autoreceptors thereby increasing norepinephrine and serotonin release, and it blocks post-synaptic 5HT-2A, 5HT-2C, and 5HT-3 serotonin receptors.(Feighner 1999) (Mianserin, an earlier analog of mirtazapine marketed in Europe, has a similar mechanism of action). Mirtazapine can improve appetite (likely through 5HT-3 and H1 antagonism) and sleep (through H1 antagonism). As expected, these immediate effects can be very beneficial in the treatment of the acutely depressed patient with poor oral intake and insomnia. Weight gain however can be a concern over the long run.

Nefazodone is a post-synaptic 5HT2 antagonist with weak serotonin and norepinephrine reuptake inhibition.(DeVane, Grothe, et al. 2002) Although nefazodone can improve sleep, and is neutral in regard to weight gain and less likely than SSRIs to cause sexual side effects, it is used much less often since it was found to produce rare (1 in 250,000 to 300,000 patient-years), but severe, hepatotoxicity.(Gelenberg 2002) **Trazodone**, a structurally similar antidepressant, is used primarily as a hypnotic (it proved to be too sedating for most patients at doses necessary for antidepressant effect). Trazodone can commonly cause orthostasis and should be used cautiously in the elderly. Priapism is a rare side effect that should be discussed with male patients before treatment.

Further Notes on the Clinical Use of Antidepressants

Clinical practice today emphasizes the use of newer ('second generation') antidepressants including bupropion, citalopram, duloxetine, escitalopram, fluoxetine, fluvoxamine, mirtazapine, nefazodone, paroxetine, sertraline, trazodone and venlafaxine. As discussed above, the older tricyclics and MAOIs are not first-line because of their greater toxicity and risk of harm from overdose. In a meta-analysis of 203 studies comparing the efficacy and side effects of these newer antidepressants, no substantial differences in effectiveness were found.(Gartlehner, Gaynes, et al. 2008) The authors recommended that antidepressants be selected on the basis of differences in expected side effects and cost (i.e. – use generic products over brand items). Another review of 117 trials concluded that sertraline had the most favorable balance among benefits, side effects, and acquisition cost.(Cipriani, Furukawa, et al. 2009)

The STAR*D study (Sequenced Treatment Alternatives to Relieve Depression), sponsored by the NIMH was a study of medications for the treatment of major depression. It produced important insights into the optimum use of pharmacotherapy for this disorder. STAR*D started with almost 4,000 heterogeneous "real world" depressed patients, who were treated by "real world" clinicians such as primary care doctors. Patients agreed to have up to 4 medication trials with the goal of achieving remission from their depression. Each trial lasted up to 14 weeks. Patients started with citalopram for the first trial. If response was unsatisfactory, they could have a switch to one of three antidepressants, or an augmentation with one of two augmenting agents. For the third trial, there were other switches or augmentations available, and finally for those still depressed and still willing to undergo the fourth trial, there was the choice of an MAOI or a combination of venlafaxine and mirtazapine. The latter has been referred to informally as 'rocket fuel' because of the four different neurotransmitter alterations that this combination is thought to induce.(McGrath, Stewart, et al. 2006) Key findings from STAR*D include the following:

- Citalopram did not work well if patients met the DSM-IV criteria for melancholic features. (McGrath, Khan, et al. 2008)

- The switches in the second trial (to another SSRI: sertraline, to bupropion, or to venlafaxine) had equal efficacy.

- The augmentations in the second trial (buspirone--discussed in anxiolytic section below--or bupropion) worked equally well.

- *Nothing* worked well in trials one or two if patients had significant anxiety symptoms along with their depression.(Fava, Rush, et al. 2008) However, a recent study with adjunctive aripiprazole (an antipsychotic discussed below) added to an SSRI found good results in patients with depression mixed with anxiety, in a post-hoc analysis.(Trivedi, Thase, et al. 2008) This needs replication in a prospectively designed study.

- In the third trial, switching to a tricyclic antidepressant worked fairly well. It might have worked better if clinicians had dosed it properly and used plasma levels to monitor adequacy of dosage.

- Adding lithium (discussed in the mood stabilizer section below) did not work as well as adding triiodothyronine in the third trial, but lithium might have done better if clinicians had dosed it properly.

- In trial 4, the MAOI did not do well compared to the venlafaxine/mirtazapine combination, but clinicians underdosed the MAOI. Unfortunately, for the few patients who improved from either treatment, early relapse was common.

As a group, STAR*D subjects were not particularly interested in psychotherapeutic treatment for their depression. Psychotherapy was available as an option in the second treatment trial, but patients could elect to drop it from the option list and most did so.(Wisniewski, Fava, et al. 2007) The modest remission rates seen in STAR*D may reflect that a major component of the improvement in depression seen in research and clinical settings comes from the non-specific, interpersonal supportive aspects of care including the therapeutic alliance. STAR*D patients might have been less susceptible to these benefits than other patients who are more invested in psychosocial treatments of their disorder. It is hoped that future studies will improve our ability to select the best treatments for each patient, psychopharmacological and psychotherapeutic, depending on their needs and preferences.

Table 1 summarizes characteristics of commonly used antidepressants.(WHO 2007; PDR 2008; Hyman, Arana, et al. 1995; Perry, Zeilmann, et al. 1994; Rosenbaum, Arana, et al. 2005; Stahl 2005; Taylor, Paton, et al. 2007)

TABLE 1. COMMONLY USED ANTIDEPRESSANTS

MEDICATION*	DOSING**	COMMENTS/ *FDA Indication*
Imipramine (TCA) (Tofranil®)	See nortriptyline, except increase gradually to 100-200 mg po qhs	Check baseline ECG; therapeutic serum level of imipramine + its metabolite desip-ramine: 175-350 ng/mL; TCA most commonly used in comparative anxiety studies; CYP1A2, CYP2D6 substrate. *Depression/Temporary adjunct in childhood enuresis in patients greater or equal to 6 years of age*

Amitriptyline (TCA) (Elavil®)	See nortriptyline, except increase gradually to 100-200 mg po qhs	Check baseline ECG; possible therapeutic serum level of amitriptyline + its metabolite nortriptyline: 93-140 ng/mL; frequently used in low doses for chronic pain; most anticholinergic TCA; TCA with most overall adverse effects; CYP1A2, CYP2D6 substrate. On WHO Essential Medicines List for depressive disorders. *Depression*
Clomipramine (TCA) (Anafranil®)	See nortriptyline, except increase gradually to 100-200 mg po qhs	Check baseline ECG; most serotonergic TCA; CYP1A2, CYP2D6 substrate. On WHO Essential Medicines List for OCD and panic attacks. *OCD*

Doxepin (TCA) (Sinequan®, Adapin®)	See nortriptyline, except increase gradually to 100-200 mg po qhs	Check baseline ECG; very sedating TCA, usually used in low doses (e.g. 10-25 mg) as adjunct for insomnia; CYP2D6 substrate. *Depression/Anxiety*
Desipramine (TCA) (Norpramin®)	See nortriptyline, except give in am and/or in divided doses, gradually increase to 100-200 mg/day	Check baseline ECG; serum therapeutic level of desipramine: greater than 115 ng/mL; least sedating (possibly activating) TCA; most noradrenergic TCA; CYP2D6 substrate. *Depression*

Nortriptyline (TCA) (Aventyl®, Pamelor®)	Start: 10-25 mg po qhs and increase by 10-25 mg every 2 days until 50-150 mg/day in divided doses then check serum level	Check baseline ECG; therapeutic serum level of nortriptyline: 58-148 ng/mL (TCA with most defined serum level—inverted U dose-response curve); TCA with least postural hypotension so best for use in elderly; CYP2D6 substrate. *Depression*
Phenelzine (MAOI) (Nardil®)	Start: 15 mg po bid and increase weekly by 15 mg/day to 45-60 mg/day	Nonselective MAOI; dangerous medication and food interactions (see package insert). *Atypical and other depressions not responsive to other antidepressants*
Tranylcypromine (MAOI) (Parnate®)	Start: 10 mg po bid and increase weekly by 10 mg/day to 30-60 mg/day	Nonselective MAOI; dangerous food and drug interactions (see package insert). *MDD without melancholia*

Transdermal Selegiline (MAOI) (Emsam®)	Start: 6 mg trans-dermal q day then increase by 3 mg patches as needed to max of 12 mg/day	Selective MAO-B inhibitor; at 6 mg dose may not need diet restrictions (but perhaps with less antidepres-sant effect), but at higher doses a nonselective MAOI and needs diet re-strictions; danger-ous food and drug interactions (see package insert). *MDD*
Fluoxetine (SSRI) (Prozac®, Prozac Weekly®, Sara-fem®)	For fluoxetine, Prozac: Start: 5-20 mg po q am then hold at 20 mg for 4 weeks then increase by 20 mg every 4 weeks as tolerated, stop if no improvement after 4 weeks at 60 mg/day	SSRI with longest ½ life, metabolite norfluoxetine with even longer ½ life; works a little slower than other an-tidepressants; inhibits CYP2C9, CYP2D6, CYP3A4. On WHO Essential Medicines List for depressive disorders. *MDD/OCD/ PMDD/Bulimia/ Panic Disorder*

Paroxetine (SSRI) (Paxil®, Paxil CR®)	For paroxetine, Paxil: Start: 10-20 mg po qhs and increase in 2-4 weeks to 30-40 mg/day as tolerated	SSRI most likely to cause discontinuation symptoms; SSRI most associated with treatment-emergent suicidality; only SSRI that produces weight gain; may have most sexual side effects; inhibits CYP2D6. *MDD/OCD/ Panic Disorder/ Social anxiety disorder PTSD/GAD/ PMDD*
Sertraline (SSRI) (Zoloft®)	Start: 25-50 mg po q day and maintain for 2-4 weeks, increase by 50 mg/day every 4 weeks if needed, maximum 200 mg/day but unclear if more helpful than 100 mg/day	Less enzymatic inhibition than fluoxetine, paroxetine, and fluvoxamine (although may increase lamotrigine levels); well-tolerated SSRI; may have the most favorable balance among benefits, side effects, and cost. *MDD/PMDD/ Panic disorder PTSD/Social anxiety disorder OCD*

Fluvoxamine (SSRI) (Luvox®, Luvox CR®)	For fluvoxamine, Luvox: Start: 25 mg po bid and increase in 4 days to 100 mg/day in single or divided doses, may increase to 200 mg/day in divided dose if needed	Primarily used for OCD in U.S. due to initial application to FDA for this indication; inhibits CYP1A2, CYP2C9, CYP2C19, CYP3A4. *OCD* *Social anxiety disorder*
Citalopram (SSRI) (Celexa®)	Start: 10-20 mg po q day and increase to 40 mg/day in 7 days, (20 mg/day may equal placebo in some studies), increase to 60 mg/day if necessary but unclear if more helpful	Least likely SSRI (along with escitalopram) to cause medication interactions; well tolerated overall. *Depression*
Escitalopram (SSRI) (Lexapro®)	Start: 10 mg po q day, increase to 20 mg/day in 2 weeks if necessary	S-citalopram; well tolerated; low risk of medication interactions; a non-generic SSRI therefore expensive; comparison with citalopram showed about 15% better efficacy with escitalopram but this may have been an artifact of doses used. *MDD/GAD*

Venlafaxine (SNRI) (Effexor®, Effexor XR®)	For venlafaxine, Effexor: Start: 37.5 mg po q day for 4 days then increase to 75 mg daily, then add 75mg/day every week until 225 mg/day	Check baseline blood pressure, then every 3-6 months; an SSRI at low doses; >150 mg needed for norepinephrine effect—but increases blood pressure at these higher doses; high risk of discontinuation syndrome; low risk of enzyme inhibition; venlafaxine is the only generic and inexpensive SNRI. *Depression/GAD/ Social anxiety disorder/Panic disorder*
Duloxetine (SNRI) (Cymbalta®)	Start: 40 mg/day in single or divided doses, increase to 60 mg/day in divided doses after 3-7 days, max 120 mg/day but unclear if more helpful	Check baseline blood pressure, then every 3-6 months; serotonergic and noradrenergic effects at all doses; modest inhibition of CYP2D6, CYP1A2. *MDD/GAD/ Diabetic peripheral neuropathy/Fibromyalgia*

Bupropion (Wellbutrin®, Wellbutrin SR®, Wellbutrin XL®, Zyban®)	For bupropion, Wellbutrin, Zyban: Start: 75 mg po bid or 100-150 mg q am and increase to 100-150 mg bid (2nd dose in afternoon) after 4-7 days, different dosing for different formulations	Contraindicated in patients with history of seizure, eating disorder or if otherwise at high seizure risk; least likely to cause sexual side effects or weight gain; modest inhibition of CYP2D6. *MDD/Prevention of MDE in patients with seasonal affective disorder/Aid to smoking cessation treatment*
Mirtazapine (Remeron®)	Start: 7.5-15 mg po qhs and increase to 30 mg q hs after 4-7 days, max 45 mg po qhs	Improves appetite and sleep as early side effects; low risk of medication interactions; less sexual side effects than SSRIs; may be more sedating at lower doses; may work faster than other antidepressants. *MDD*

Trazodone (Desyrel®)	For insomnia only: Start: 25 mg po qhs, if needed increase to 50 mg, then can increase by 50 mg increments up to 200 mg at bedtime	Used primarily for insomnia; may cause orthostasis, priapism; no longer used as antidepressant but when used as antidepressant dose was 400 mg daily; CYP3A4 substrate.
		Depression

*Generic and U.S. brand name(s). **Dosing should be adjusted downwards ('start low, go slow' strategy) for the elderly and/or the medically compromised. Abbreviations: bid-(bis in die) twice a day; CYP-Cytochrome P450 enzyme; FDA-Food and Drug Administration; GAD-Generalized Anxiety Disorder; MAOI-Monoamine Oxidase Inhibitor; MAO-B-Monoamine Oxidase Inhibitor, B subtype; MDD-Major Depressive Disorder; MDE-Major Depressive Episode; mg-milligram; ng/mL-nanogram per milliliter; OCD-Obsessive Compulsive Disorder; PMDD-Pre-menstrual Dysphoric Disorder; po-(per os) orally; PTSD-Post-traumatic Stress Disorder; q-(quaque) every; qhs-(quaque hora somni) at bedtime; SNRI-Serotonin Norepinephrine Reuptake Inhibitor; SSRI-Selective Serotonin Reuptake Inhibitor; TCA-Tricyclic Antidepressant; WHO-World Health Organization.

ANXIOLYTICS

The pharmacological treatment of anxiety symptoms is both simple and complicated. On the one hand, medications such as benzodiazepines and barbiturates can have a relatively immediate effect on distressing anxiety symptoms. On the other hand, the use of such medications carries the risk of cognitive impairment, physical dependence, as well as the risk of psychological dependence and inappropriate use in some patients.

It is not clear that episodic anxiety that is associated with situational stressors should be treated with medications. Anxiety *per se* may be a normal response to distressing events and a signal that may enhance a person's motivation to address these real-life events. As such, it may be better understood and addressed through psychotherapy rather than pharmacologically. Medical students and physicians should be aware of cultural (and managed care) pressures that push for 'popping a pill' rather than addressing the underlying causes of the patient's anxiety and improving coping strategies.

Arash Ansari, M.D. and David N. Osser, M.D.

Anxiety disorders are characterized by persisting patterns of anxiety symptoms impairing functioning. Examples include panic disorder, social anxiety disorder, post-traumatic stress disorder, obsessive-compulsive disorder and generalized anxiety disorder. The first-line medication treatments for most anxiety disorders are SSRIs (or other antidepressants with serotonergic effects—see Table 1). A time period of several weeks may be necessary before clear response. During this time, anxiolytics with more immediate effects (e.g. benzodiazepines) may be used for early symptom control.

Benzodiazepines

Benzodiazepines were first developed in the 1960s and are now the most commonly used anxiolytics used in the United States. **Diazepam**, **clonazepam**, **chlordiazepoxide**, **temazepam**, **oxazepam**, **lorazepam**, and **alprazolam** are examples of benzodiazepines. Their mechanism of action is through their binding on γ-aminobutyric acid (GABA) receptors.(Nutt and Malizia 2001) GABA is the primary inhibitory neurotransmitter in the brain. Benzodiazepines bind to one type of GABA receptor $(GABA_A)$ thereby increasing the receptor's affinity for GABA. Increased GABA effect then increases the frequency of chloride channel openings allowing this ion's influx into the cell which in turn decreases normal cell firing. The benzodiazepine binding site is composed of multiple subunits; binding to the alpha-1 subunit may explain sedative effects of benzodiazepines whereas alpha-2 subunit binding may be needed for anxiolytic effects.(Nestler, Hyman, et al. 2009)

Benzodiazepines are associated with multiple adverse effects. They are sedating, can impair concentration, memory,(Buffett-Jerrott and Stewart 2002) coordination, can lead to falls in the elderly(Wagner, Zhang, et al. 2004) (especially at initiation of treatment and after dose increases), and can cause respiratory depression. The choice of which benzodiazepine to use is often based on the pharmacokinetics of each drug. Diazepam, chlordiazepoxide, and clonazepam have relatively long half-lives. Additionally, diazepam and chlordiazepoxide are significantly hepatically metabolized and have multiple active metabolites; the use of these medications in hepatically compromised patients is therefore problematic. In medically ill patients, and in the elderly, benzodiazepines with short half-lives, such as lorazepam and oxazepam, are preferred, especially when the risk of respiratory depression is a concern (e.g. patients with chronic obstructive pulmonary disease). Alprazolam has a shorter half-life than lorazepam and oxazepam. It is, however, also associated with significant rebound anxiety because of the rapid drop from peak serum level after each dose. Despite its current widespread use, alprazolam should generally be avoided in patients who may require frequent or daily administration of an anxiolytic drug.

Perhaps the greatest drawback of benzodiazepines, however, is that they can lead to abuse and/or dependence. Physical dependence is characterized by increased tolerance to these drugs and the development of significant withdrawal symptoms upon discontinuation; this occurs with long term and/or high dose use of benzodiazepines and is not necessarily a sign of misuse or addiction (although patients should be made aware of the need for very gradual taper of these medications if used long term). Addiction, on the other hand, which may include elements of physical dependence, is characterized by maladaptive behavioral changes leading to medication misuse. Benzodiazepines (along with barbiturates discussed below) are controlled substances which should be prescribed judiciously and cautiously and only when adequate follow-up is available to ensure appropriate use. It should be noted that adequate follow-up is not often feasible in primary care settings. Benzodiazepines should generally be avoided in any patient with a history of substance or alcohol abuse: most benzodiazepine abuse or dependence occurs in these individuals. Admittedly, there are circumstances in which a patient with a history of substance abuse may require benzodiazepines (for example a patient with a severe debilitating panic disorder who has been refractory to all other non-benzodiazepine medications)—these circumstances, however, should be considered infrequent.

Barbiturates

Developed in the 1940's and 50's, barbiturates are now rarely used for the treatment of anxiety due to a higher risk of dependence and dangerousness in overdose when compared to benzodiazepines. Whereas benzodiazepine binding increases the receptor's affinity for GABA and indirectly affects chloride channels, barbiturates (and alcohol), binding on a different site on $GABA_A$ receptors, can increase chloride influx into neurons even when GABA is not present. **Chloral hydrate**, a weak barbiturate, is still occasionally used in certain settings for the treatment of refractory insomnia or for sedation prior to anxiety provoking medical studies (e.g. MRI). Other barbiturates such as **phenobarbital**, **pentobarbital**, and **butalbital,** are still commonly used in other areas of medicine for treatment of conditions (e.g. seizure disorder, pain) other than anxiety disorders.

Medicines without Abuse Potential Used for the Treatment of Anxiety

Buspirone is a partial 5HT1A agonist (primarily on autoreceptors) causing decreased serotonin release from serotonergic neurons. It is not clear why this action would help decrease anxiety; any effect is presumed to be due to 'downstream' adaptations over several weeks. Buspirone has no effect on GABA receptors and as such cannot immediately replace benzodiazepines. It has no immediate anxiolytic effects. However, it has no potential for abuse, and does not impair cognition or motor coordination. Side effects, however, may include headache, insomnia, jitteriness and nausea. It is indicated for the treatment of Generalized Anxiety Disorder.

Propranolol is a beta-adrenergic antagonist. Although it is primarily used medically for its effect on heart rate and blood pressure, its 'off-label' use in psychiatry is based on its ability to reduce overall sympathetic activation. It is particularly helpful in circumstances where a sympathetic reaction to an anxiety-provoking stimulus can occur, such as in instances of performance anxiety. Musicians, for example, may take a dose one hour prior to their appearance on stage. Propranolol can decrease somatic manifestations of anxiety such as tremulousness and tachycardia. It does not, however, help alleviate symptoms associated with generalized social phobia or generalized anxiety disorder. Propranolol should be avoided if the patient has congestive heart failure or significant asthma. Despite earlier concerns that beta-blockers may cause depression(Waal 1967) this is not supported by more recent studies.(Ko, Hebert, et al. 2002)

Clonidine, initially developed as an antihypertensive, is an alpha-2-adrenergic autoreceptor agonist which serves to decrease sympathetic drive in the locus ceruleus. It can decrease hyperarousal in patients with post-traumatic stress disorder(Boehnlein and Kinzie 2007) and other conditions associated with autonomic hyperactivity (e.g. rebound hyperactivity in opioid withdrawal states).

Prazosin is an alpha-1-adrenergic receptor antagonist. Like clonidine, it is an antihypertensive which can decrease anxiety symptoms associated with post- traumatic states. It has no sedative properties but it can help decrease PTSD symptoms during the day and decrease associated sleep disturbances and nightmares at night.(Miller 2008; Raskind, Peskind, et al. 2007; Taylor, Lowe, et al. 2006; Taylor, Martin, et al. 2008)

Hydroxyzine is an antihistamine with less affinity for muscarinic and alpha-1-adrenergic receptors than other antihistamines. Because it does not cause dependence and has no abuse potential, it is useful for treating anxiety symptoms in patients with a history of substance abuse. Also, it has been shown to have efficacy in the treatment of generalized anxiety disorder.(Llorca, Spadone, et al. 2002; Lader and Scotto 1998; Ferreri, and Hantouche, et al. 1994)

Newer Hypnotics

Zolpidem, **zaleplon**, and **eszopiclone** (enantiomer of zopiclone which is available in Europe) are newly developed non-benzodiazepine hypnotics that bind to alpha-1 subunits on the benzodiazepine binding site on GABA receptors.(Sanger 2004) These "z-drugs" cause sedation but lack anxiolytic effects despite some cross-reactivity with benzodiazepines. Although their abuse potential is purportedly less than benzodiazepines, they are not free from the risk of dependence and withdrawal symptoms upon discontinuation.(Cubala and Landowski 2007; Liappas, Malitas, et al. 2003; Sethi and Khandelwal 2005; Victorri-Vigneau, Dailly, et al. 2007) There is insufficient evidence that these hypnotics are either safer or more effective than benzodiazepines, yet, due to physician misconceptions, they have been widely used when more cost-effective treatments should be considered.(Siriwardena, Qureshi, et al. 2006; Glass, Lanctot, et al. 2005) Zolpidem is now generic, but zaleplon and eszopiclone are still brand products and are expensive. **Ramelteon** is a new hypnotic that is a melatonin receptor agonist which may have modest efficacy in shortening sleep latency. (Sateia, Kirby-Long, et al. 2008; Roth, Seiden, et al. 2006) It is well-tolerated(Johnson, Suess, et al. 2006) but again there are no studies to suggest it should be favored over more cost-effective alternatives. When considering treatment of chronic insomnia, physicians should not overlook the benefits that may be derived from nonpharmacological, e.g. behavioral, therapies. (Sivertsen, Omvik, et al. 2006)

Table 2 summarizes the characteristics of selected non-antidepressant medications for the treatment of anxiety and insomnia.(WHO 2007; PDR 2008; Hyman, Arana, et al. 1995; Rosenbaum, Arana, et al. 2005; Stahl 2005; Taylor, Paton, et al. 2007) Antidepressants used in the treatment of anxiety disorders are listed in Table 1.

TABLE 2. SELECTED NON-ANTIDEPRESSANT MEDICATIONS FOR ANXIETY AND INSOMNIA

MEDICATION*	DOSING**	COMMENTS/ *FDA Indication*
Clonazepam (Benzodiazepine) (Klonopin®)	Start: 0.5 mg po bid for panic disorder, increase as needed, use lowest effective dose. Equivalence: 0.25 mg equals lorazepam 1 mg	Benzodiazepine with convenient pharmacokinetics for the treatment of panic disorder (30-50 hours half-life); CYP3A4 substrate. *Panic disorder/Specific seizure disorders (see package insert)*

Diazepam (Benzo-diazepine) (Valium®, Diastat®, Diazepam Injection®)	For oral diazepam, Valium: Start: 2 mg po bid-tid, increase as needed, use lowest effective dose. Equivalence: 5 mg equals lorazepam 1 mg	Rapid onset of action followed by rapid distribution to lipid compartment, long elimination half-life. On WHO Essential Medicines List for generalized anxiety and sleep disorders, and as an anticonvulsant. *Anxiety disorders and symptoms/ Alcohol withdrawal symptoms/Adjunctive treatment for convulsive disorders/ Adjunctive therapy in skeletal muscle spasms*
Chlordiazepoxide (Benzodiazepine) (Librium®)	Start: 10 mg po tid-qid, increase as needed, use lowest effective dose. Equivalence: 25 mg equals lorazepam 1 mg	Frequently used in inpatient detoxification for alcohol withdrawal symptoms when there is no hepatic dysfunction; multiple psychoactive metabolites. *Anxiety disorders and symptoms/Alcohol withdrawal symptoms/Preoperative anxiety and apprehension*

Oxazepam (Benzo-diazepine) (Serax®)	Start: 10 mg po tid, increase as needed, use lowest effective dose. Equivalence: 15 mg equals lorazepam 1 mg	Used in inpatient detoxification when hepatic impairment is present; slowest onset of action among benzodiazepines. *Anxiety/Alcohol withdrawal*
Lorazepam (Benzo-diazepine) (Ativan®, Ativan Injection®)	For oral lorazepam, Ativan: Start: 0.5 mg po bid-tid, increase as needed, use lowest effective dose	Most widely used in inpatient setting for 'as needed' treatment of anxiety, agitation, and withdrawal states; only benzodiazepine available IM. *Anxiety Disorders and symptoms/Status epilepticus (for injection)/Preanesthetic medication for adults (for injection)*
Alprazolam (Ben-zodiazepine) (Xanax®, Xanax XR®, Niravam®)	For alprazolam, Xanax, Niravam: Usual starting dose is 0.25 mg po tid, change to other benzodiazepine if ongoing treatment is needed. Equivalence: 0.5 mg equals loraze-pam 1 mg	Most addictive benzodiazepine; infrequent 'as needed' use may be appropriate; CYP3A4 substrate. *Anxiety disorders and symptoms/Panic disorder*

Buspirone (Atypical anxiolytic) (Buspar®)	Start: 5 mg po bid-tid; increase every 2-3 days by 5-10 mg to 30-40 mg in two divided doses, maximum dose is 60 mg/day	Alcoholics with anxiety may require near maximum doses; CYP3A4 substrate. *Anxiety disorders and symptoms corresponding to GAD*
Propranolol (Beta-blocker) (Inderal®, Inderal LA, Innopran XL®)	For propranolol, Inderal: Start: test dose of 10 mg po, then up to 40 mg 0.5-1 hour before anxiety provoking event	Also used in psychiatry to treat akathisia, lithium-induced tremor, and clozapine-induced tachycardia. *Migraine prophylaxis/ Hypertension/ Other cardiac conditions (see package insert)*
Clonidine (Antihypertensive) (Catapres®, Catapres-TTS®)	For oral clonidine, Catapres: Start: 0.1 mg po bid-tid or qhs, increase as needed and tolerated	Also used for opiate withdrawal. *Hypertension*

Prazosin (Antihypertensive) (Minipress®)	Start: 1 mg po qhs, after 3 days increase to 2 mg po qhs, after 4 more days increase to 4 mg po qhs, if no response after 7 days then increase to 6 mg po qhs, after another 7 days increase to 4 mg po at 3 pm and 6 mg po qhs	Helpful for insomnia and nightmares associated with PTSD. *Hypertension*
Hydroxyzine (Antihistamine) (Atarax®, Vistaril®)	Start: 10-12.5 mg po bid and 20-25 mg po qhs, increase as needed and tolerated	May also have analgesic effects; antiemetic, antihistamine, may help with insomnia. *Anxiety symptoms/ Multiple other indications(see package insert)*
Zolpidem (Hypnotic) (Ambien®, Ambien-CR®)	For Ambien: Start: 5 mg po qhs, may increase to 10 mg po qhs if needed	Rapid onset; reported cases of amnesia; sertraline may increase serum level. *Short-term treatment of insomnia characterized by difficulties with sleep initiation*

Zaleplon (Hypnotic) (Sonata®)	Start: 5 mg po qhs, may increase to 10 mg qhs, maximum 20 mg po qhs	Amnesia may occur as it does with benzodiazepines; ultra-short half-life; expensive; CYP3A4 substrate. *Short-term treatment of insomnia*
Eszopiclone (Hypnotic) (Lunesta®)	Start: 1-2 mg po qhs, maximum 3 mg po qhs	Amnesia may occur; similar dependence potential to diazepam; expensive with no advantages; CYP3A4 substrate. *Treatment of insomnia*
Ramelteon (Melatonin receptor agonist) (Rozerem®)	8 mg po within 30 minutes of bedtime, do not take with fatty meal	No DEA restriction; very short half-life; expensive; CYP1A2, CYP2C9, and CYP3A4 substrate. *Treatment of insomnia characterized by difficulty with sleep onset*

*Generic and U.S. brand name(s). **Dosing should be adjusted downwards ('start low, go slow' strategy) for the elderly and/or the medically compromised. Abbreviations: bid-(bis in die) twice a day; CYP-Cytochrome P450 enzyme; DEA-Drug Enforcement Administration; FDA-Food and Drug Administration; IM-intramuscular; mg-milligram; po-(per os) orally; PTSD-Post-traumatic Stress Disorder; qhs-(quaque hora somni) at bedtime; qid-(quater in die) four times a day; tid-(ter in die) three times a day; WHO-World Health Organization.

ANTIPSYCHOTICS

First Generation Antipsychotics (FGAs)

The first antipsychotic, chlorpromazine, was developed in the 1950s.(Meyer and Simpson 1997) Subsequently other antipsychotics were developed that share similarities in their mechanisms of action and in their side effect profiles. **Chlorpromazine, thioridazine, perphenazine, molindone, thiothixene, pimozide, fluphenazine,** and **haloperidol** are examples of these medications which are now characterized as 'first generation antipsychotics' (FGAs). Alternative names for this class of medications include: 'neuroleptics' (for their propensity to cause adverse neurological effects), 'major tranquilizers' (as opposed to the later designated 'minor tranquilizers' such as benzodiazepines and barbiturates), 'typical' antipsychotics, and 'conventional' antipsychotics.

All first generation antipsychotics are believed to exert their antipsychotic effects through post-synaptic

D2 dopamine receptor antagonism, thereby reducing the effect of endogenous dopamine released by presynaptic dopaminergic neurons.(Nestler, Hyman, et al. 2009) Dopaminergic neurons originate from 3 distinct nuclei. One group of dopaminergic neurons projects from the ventral tegmental area of the midbrain to the nucleus accumbens, cingulate cortex and prefrontal cortex (the mesolimbic and mesocortical tracks); these affect emotions and cognition and as such are the targets for the therapeutic effects of antipsychotic drugs. Dopaminergic neurons also arise from the substantia nigra and project to the striatum (the nigrostriatal track); these are implicated in the neurological side effects of antipsychotics. And finally, hypothalamic dopaminergic neurons project to the pituitary gland and serve to regulate the release of prolactin (the tuberoinfundibular track); disruption of this system with D2 blockade can result in hyperprolactinemia associated with the use of antipsychotics. Overall, in terms of clinical use, it has been shown that the optimal D2 receptor occupancy level for maximizing antipsychotic effect while minimizing adverse effects is 60-70%.(Farde, Nordstrom, et al. 1992)

FGAs have traditionally been divided into low potency (e.g. chlorpromazine, thioridazine), mid-potency (e.g. perphenazine, molindone, thiothixene), and high potency (e.g. haloperidol, fluphenazine) antipsychotics, based on the number of milligrams of each drug needed to show comparable efficacy. For example, chlorpromazine 300 mg (low potency) may have the same therapeutic effect as perphenazine 24 mg (mid-potency) and haloperidol 6 mg (high potency). FGAs are often listed as a spectrum from low to high potency. The low potency antipsychotics usually exhibit TCA-like side effects such as anticholinergic, antihistaminic, and orthostatic effects (see section on antidepressants) but have a lower risk of causing acute neurological side effects such as acute muscle dystonias. At the other end of the spectrum, high potency FGAs have lower risks of TCA-like adverse effects but a much higher risk of causing acute dystonias. Mid-potency antipsychotics share all these side effects but less so than those of either pole. Physicians should become familiar with using at least one antipsychotic from each potency class in order to be able to match the side effect profile to the patient's pre-existing vulnerabilities.

When dosing FGAs, it is important to consider that although in most efficacy studies the presumed therapeutic dose of haloperidol is 10 mg/day, the ideal dose may be much lower: haloperidol 2 mg/day in neuroleptic-naïve patients, and 4 mg in non-neuroleptic-naïve patients may be sufficient to produce a 'neuroleptic threshold'—the dose at which cogwheel rigidity, a sign of sufficient D2 receptor blockade, first appears.(McEvoy, Stiller, et al. 1986)

The propensity of FGAs to cause neurological symptoms such as acute muscle dystonias, parkinsonism, akathisia, and tardive dyskinesia significantly limits their use in current practice. *Acute dystonias*, which are more likely to occur if the patient is young, male, has a history of substance abuse and/or a prior history of dystonias, is primarily seen in patients taking FGAs although it can also occur in patients with any antipsychotic with significant D2 receptor antagonism (see risperidone below). Use of anticholinergic medications, such as benztropine or diphenhydramine (or promethazine used outside the U.S.) can decrease the occurrence of early dystonias. *Parkinsonism*, characterized by bradykinesia, tremor, rigidity, and masked facies, can develop after 1-4 weeks of treatment with FGAs. Anticholinergic medications or a dopamine releasing agent such as amantadine may be helpful, although changing the antipsychotic may be required. *Akathisia*, which is an unpleasant subjective sense of inner restlessness relieved by movement, is also commonly seen in patients treated with FGAs. Identifying akathisia as a cause of agitation (or even worsening psychosis or suicidality) is important because treatment would include decreasing, rather than increasing, antipsychotic dose. *Tardive dyskinesia* (TD), a potentially irreversible syndrome of abnormal involuntary movements, can develop with extended use of antipsychotics, especially if high doses are used for long periods of time. Patients with prolonged antipsychotic treatment, a history of affective disorders, a history of parkinsonian side effects with initial antipsychotic treatment, as well as women and the elderly, are at a higher risk for developing TD. Once tardive dyskinesia has developed, withdrawal of the antipsychotic (especially if this is precipitous) may unmask worsened abnormal movements. Resumption of antipsychotic treatment may suppress these symptoms for a period of time, but progression of the underlying movement disorder may continue. Despite trials of multiple remedies, treatments for TD are generally only partially effective.(Soares-Weiser and Fernandez 2007)

Neuroleptic malignant syndrome (NMS), is a poorly understood, rare, but potentially fatal complication of treatment with FGAs and other antipsychotics. NMS is characterized by a constellation of symptoms which may include delirium, lead-pipe rigidity, autonomic instability and high fevers. It can develop very early in the course of antipsychotic treatment. A high serum creatine phosphokinase (CPK) and elevated white blood cell count are supportive of the diagnosis of NMS. If NMS appears likely then the offending antipsychotic should be immediately discontinued. Medical hospitalization is necessary and treatment may include the use of a dopamine agonist (e.g. bromocriptine), a muscle relaxant (e.g. dantrolene), aggressive hydration, and the use of benzodiazepines if needed for behavioral agitation.(Hu and Frucht 2007) Once the patient has been medically stabilized, the offending agent should be avoided. Rechallenge may be possible two weeks after all symptoms of NMS have abated, optimally with a low potency FGA or a second generation antipsychotic.

Second Generation Antipsychotics (SGAs)

Second generation antipsychotics, also known as atypical antipsychotics, are believed to exert their antipsychotic effects through a similar mechanism of action, but have profiles of receptor activity that produce different side effects than FGAs. The first SGA to be developed was **clozapine**, which was followed by the sequential introduction of **risperidone**, **olanzapine**, **quetiapine**, **ziprasidone**, and **aripiprazole** in the U.S. (and amisulpride and zotepine among others, in other countries). Whereas FGAs were known to (1) possibly worsen (or at least only partially improve) the negative symptoms of schizophrenia and (2) cause extrapyramidal symptoms including TD, SGAs were hoped to be more effective in treating negative symptoms and less likely to cause movement disorders. It is true that, by and large, SGAs do not worsen negative symptoms of schizophrenia and have a much lower risk of causing TD.(Correll, Leucht, et al. 2004) However, questions regarding the differential effectiveness of SGAs as compared with FGAs, and their greater risks of inducing other, non-neurological, adverse effects have served to dampen the optimistic expectations initially associated with the medications. Nevertheless, in most of the world where they are available, SGAs are still considered to be the first line for treatment of psychotic disorders.

Risperidone, one of the earliest SGAs to be developed, was released in 1994 and is now generic. It is similar to FGAs in that it is a potent D2 receptor antagonist, but like many other SGAs, it is also a post-synaptic serotonin 5HT2A antagonist. This is thought to mitigate the D2 receptor-mediated neurological side effects. At doses lower than 6 mg/day (i.e. at usual therapeutic doses), risperidone carries a low risk of causing EPS; at higher doses, D2 blockade effects predominate and the risk of EPS increases significantly. EPS are usually not present at risperidone 3 mg/day—a dose at which 72% of D2 receptors are occupied.(Nyberg, Eriksson, et al. 1999) The optimal dose derived from clinical studies appears to be 3-6 mg daily.(Osser and Sigadel 2001) Although generally a well-tolerated antipsychotic, the side effects of risperidone include possible hypotension and hyperprolactinemia, and in children and adolescents it produces considerable weight gain.(Sikich, Frazier, et al. 2008) It is a hepatic CYP2D6 enzyme substrate and therefore its metabolism can be slowed by (1) inhibitors such as fluoxetine and paroxetine,(Spina, Scordo, et al. 2003) or (2) the CYP2D6 variant gene for slow metabolism, which results in a less active form of the CYP2D6 enzyme (found more often in Chinese and other East Asian individuals).(Bertilsson 1995) Another gene variant that causes "poor" metabolism is more common in Caucasians and results in severe side effects. On the positive side, risperidone has a low to medium propensity to cause adverse metabolic effects in adults (see discussion below). It may also have a somewhat more rapid onset of action compared to other second generation antipsychotics.(Osser and Sigadel 2001) The newly introduced (and therefore more costly) **paliperidone**, which is the major active metabolite of risperidone with similar efficacy and similar side effect profile, is not metabolized by CYP2D6 and is mostly renally excreted.

Olanzapine was introduced in 1996. It has less affinity for D2 receptors than risperidone and a greater affinity for 5HT2A and 5HT2C serotonin receptors. Olanzapine also has significant antihistaminic and anticholinergic effects. Although it is an effective antipsychotic for the treatment of schizophrenia (especially at doses equal or greater than 15 mg/day),(Osser and Sigadel 2001) it is (along with clozapine as discussed below) associated with a high risk of developing weight gain, insulin resistance, and hyperlipidemia (i.e. 'metabolic syndrome'). Concern about the increased morbidity and mortality associated with the metabolic syndrome has lead to a reduction of the use of olanzapine in recent years. All side effects increase when olanzapine is used at higher than recommended doses (e.g. 40 mg/day vs. the package insert maximum dose of 20 mg/day) with very little additional antipsychotic benefit.(Kinon, Volavka, et al. 2008) Liver transaminases can also become transiently elevated with olanzapine.

Quetiapine shows weak affinity at both dopamine and 5HT2 serotonin receptors, but may have similar receptor occupancy to the more potent SGAs.(Seeman 2002) It has alpha-adrenergic antagonism and antihistaminic effects, causing orthostasis and sedation, respectively. Quetiapine is less likely than olanzapine and clozapine, but more likely than most FGAs, risperidone and other SGAs, to cause metabolic side effects. Quetiapine, at low doses, is widely (and perhaps too readily) used in psychiatric practice for the treatment of insomnia and acute anxiety in a wide range of patients with personality and/or substance abuse disorders for whom benzodiazepine use may be problematic. This 'off-label' use should only be carried out after a thoughtful review of risks, benefits, and alternative treatments, especially evidence-supported treatments, for patients' diagnosed conditions. Clinicians should be aware also of recent reports of abuse and 'street value' for this medication. (Hanley and Kenna 2008) Use of quetiapine for anxiety symptoms may be more appropriate in acute care settings such as during hospitalizations. Quetiapine's effectiveness in psychotic disorders may be less than that of olanzapine and risperidone,(McCue, Waheed, et al. 2006; Suzuki, Uchida, et al. 2007) but it may have a stronger role in bipolar disorders (see section on mood stabilizers).

Ziprasidone is an SGA with moderate D2 antagonism and significant 5HT2A antagonism (i.e. a high 5HT2A/D2 ratio). Although it is not clear if it is as effective as olanzapine and risperidone in the acute treatment of schizophrenia,(McCue, Waheed, et al. 2006) it does not cause metabolic changes, and may even improve lipid profile, especially if the patient was previously on a weight gain-inducing agent.(Lieberman, Stroup, et al. 2005) A major issue with using ziprasidone is the necessity of taking it with food, or it will not be well-absorbed. A 500 calorie meal is optimal with each of the twice daily doses.(Miceli, Glue, et al. 2007)

Ziprasidone has the potential to prolong QT more than other SGAs. Although a pre-treatment ECG is not required, those who are deemed, based on history or age, to be at higher cardiac risk would benefit from an ECG (and medical consultation if appropriate) before starting ziprasidone. Electrolyte disturbances such as hypomagnesemia and hypokalemia should be corrected. Other QT prolonging medications should not be used in combination with ziprasidone. Despite concerns regarding this effect, post-marketing studies (e.g. CATIE)(Lieberman, Stroup, et al. 2005) did not show any clinically significant QT prolongation with ziprasidone use.

Medical students and physicians should be aware that all antipsychotics (with the possible exception of aripiprazole discussed below) can affect cardiac conduction, potentially delaying conduction enough to lead to fatal arrhythmias. There is an association between the use of antipsychotics (as well as tricyclic antidepressants) and sudden cardiac death. (Ray, Chung, et al. 2009; Ray, Meredith, et al. 2004; Straus, Bleumink, et al. 2004) As discussed in the section on antidepressants, prolonged QTc is associated with torsades de pointes, a potentially fatal arrhythmia. The QT interval includes both the QRS interval as well as the ST segment. Whereas TCAs and some FGAs with tricyclic structure (e.g. chlorpromazine) lengthen the QRS interval by interfering with sodium channels and depolarization, most other antipsychotics, including SGAs, can affect potassium channels and the repolarization phase.(Glassman and Bigger 2001) Both effects would be reflected in the QT interval. Although it is not clear if QT prolongation is actually a reliable indicator of the risk of torsades, measuring this interval is the simplest way to estimate this risk.(Shah 2005) In addition to ziprasidone, the FGAs thioridazine, **mesoridazine**, pimozide, and **droperidol** are among the antipsychotics with the highest propensity to prolong the QT interval.(Fayek, Kingsbury, et al. 2001)

Aripiprazole, in contrast to other SGAs, is a high affinity partial agonist at the D2 receptor.(Mamo, Graff, et al. 2007) It is postulated that aripiprazole decreases overall dopamine effect (Stahl 2008) in dopamine rich environments (e.g. in mesolimbic pathways--thereby ameliorating psychosis), and increases dopamine effect in dopamine depleted environments (e.g. in mesocortical pathways to the prefrontal cortex--thereby improving negative symptoms such as social withdrawal).(Stahl 2008) At therapeutic doses it highly saturates the targeted dopamine receptors and shows very slow dissociation from the receptors upon discontinuation.(Goff 2008; Grunder, Fellows, et al. 2008) It also shows moderate 5HT2A and 5HT2C antagonism. On the other hand, it is free from anticholinergic and significant antihistaminic effects. More importantly, it may not have significant cardiac or metabolic effects.(El-Sayeh, Morganti, et al. 2006)

Although it is less likely to cause EPS in general, it has been observed in practice to cause akathisia more readily than other SGAs. This side effect may be more common if the patient was recently on a strong D2 antagonist such as an FGA or risperidone and consequently has an up-regulated or hypersensitive population of D2 receptors. (Raja 2007) Slow dose titration and/or combination with a benzodiazepine may be necessary to reduce the risk of akathisia.

Aripiprazole at 15 mg/day may be more efficacious than 30 mg/day in schizophrenia, although full response may take longer than with a comparable dose of haloperidol.(Kane, Carson, et al. 2002) Higher doses (e.g. 30 mg/day) may be more useful in treatment-resistant schizophrenia.(Kane, Meltzer, et al. 2007) Relapse rates may be somewhat higher with aripiprazole than with other SGAs.(Pigott, Carson, et al. 2003)

Clozapine, the first and in some respects the most impressive of the second generation antipsychotics, binds weakly at the D2 receptor (although with relatively greater net antagonism at D3 and D4 dopamine receptors) and has moderate affinity for 5HT2A and 5HT2C receptors. It is often effective when other antipsychotics are not,(Lewis, Barnes, et al. 2006) and appears to have antisuicidal effects in patients with schizophrenia or schizoaffective disorder.(Meltzer, Alphs, et al. 2003) However, because of the risk of agranulocytosis, it is reserved primarily for schizophrenic and schizoaffective patients who have failed to respond adequately to at least two other antipsychotics. Strict monitoring and initially weekly and then biweekly blood draws are required to monitor white cell count. The clinician should consult the package insert and strictly follow the monitoring guidelines. Clozapine should not be combined with other medications (e.g. carbamazepine) that may also cause leukopenia.

Clozapine can also cause multiple other adverse effects, which include (but are not limited to) an increased risk of seizures, rare myocarditis, eosinophilia, anticholinergic and antihistaminic effects, orthostasis, weight gain and adverse metabolic effects.(Lamberti, Olson, et al. 2006)

Among the SGAs, clozapine and olanzapine are the most likely (and aripiprazole and ziprasidone are the least likely) to cause adverse metabolic effects. These would include weight gain, hyperglycemia and diabetes (with or without weight gain) and hyperlipidemia. (ADA 2004) A 2-3 kilogram weight gain within the first 3 weeks of treatment often predicts the risk of significant weight gain over the long term.(Lipkovich, Citrome, et al. 2006) Decreased insulin secretion and increased triglycerides (i.e. the lipids most affected by SGAs)(Osser, Najarian, et al. 1999) can also be seen within 1-2 weeks of treatment.(Chiu, Chen, et al. 2006) Treatment with the hypoglycemic medication metformin, especially if combined with lifestyle changes, may reduce antipsychotic-induced weight gain.(Baptista, Rangel, et al. 2007; Wu, Zhao, et al. 2008)

Prior to starting olanzapine or clozapine, measurements of baseline weight, serum glucose, and lipid profile should be obtained. If the patient has pre-existing diabetes, other antispychotics should be considered. Once treatment is initiated, serum glucose and weight should be monitored and if glucose levels become elevated, a glucose tolerance test—which can predict up to 96% of patients who would develop diabetes—should be done. (van Winkel, De Hert, et al. 2006) If metabolic problems do arise during treatment, switching to another antipsychotic should be considered.

Clozapine may cause orthostatic hypotension. Patients who are elderly, have cardiac histories, or who are taking antihypertensives are at higher risk for this side effect. Clozapine should be increased gradually after treatment is initiated (starting at 12.5 mg/day and increasing the dose by 25 mg daily as tolerated). Usually, patients adjust and become tolerant to the hypotensive effects of this medication. However, this tolerance may not last longer than 48 hours. If a patient discontinues clozapine therapy for more than 48 hours, treatment should be restarted with a 12.5 mg dose. After that, the dose may be more quickly raised to the previous dose as tolerated. It is important for the physician who may be admitting a psychiatric patient to the medical or surgical ward of the hospital to stop and think before continuing clozapine at its prior dose: recent compliance needs to be verified first.

Given the complicated nature of clozapine treatment, the clinician should refer to a more in depth discussion of this drug before use.(Phansalkar and Osser 2009; Phansalkar and Osser 2009)

Long-Acting Injectable Antipsychotics

In the United States, 3 antipsychotics are available for long-acting (i.e. depot) intramuscular administration: **haloperidol decanoate**, **fluphenazine decanoate**, and **long-acting injectable risperidone**. **Olanzapine**

long-acting injection is expected to be approved soon. These long-acting formulations are options for patients who are frequently non-adherent to the prescribed oral medication.(Olfson, Marcus, et al. 2007) A brief trial of the antipsychotic in oral form is first prescribed to assess patients' response to, and tolerance of, the selected agent. Every four week injections of haloperidol decanoate or biweekly injections of long-acting fluphenazine or risperidone are then continued while the oral agent is gradually tapered. Four to five repeated injection cycles of the selected antipsychotic may be necessary to achieve steady state before oral medications should be completely withdrawn.(Osser and Sigadel 2001) Patients who adhere poorly to oral medications in real-world public sector settings are generally non-adherent to depot antipsychotics as well.(Olfson, Marcus, et al. 2007) These formulations seem to work best in research subjects and in other populations of relatively cooperative and less treatment-resistant patients. On the other hand, once steady state is achieved, depots do have the advantage that if the patient discontinues treatment, the antipsychotic effect can continue for up to several months after the last received dose.

Antipsychotics for Behavioral Control

Both SGAs and FGAs are used in psychiatric practice to treat behavioral agitation. In acutely psychotic and/ or manic patients, FGAs, such as oral or intramuscular haloperidol (often combined with lorazepam and/or benztropine to decrease the risk of acute dystonias), remain the mainstay of treatment.(Ansari, Osser, et al.

2009, Osser and Sigadel 2001) Newer antipsychotics, such as olanzapine, ziprasidone, and aripiprazole are also available in short-acting intramuscular form but they are expensive and seem to have no advantage when compared with the combination therapy noted above.(Satterthwaite, Wolf, et al. 2008) When considering antipsychotics for behavioral agitation, clinicians should be advised not to use (1) intramuscular droperidol due to high risk of QT prolongation, (2) intramuscular chlorpromazine due to risk of severe hypotension, (3) intramuscular ziprasidone if the patient is taking other medications that can also prolong QT, or (4) intramuscular olanzapine in combination with lorazepam or other benzodiazepines due to the risk of hypotension.(Zacher and Roche-Desilets 2005)

The use of antipsychotics for the treatment of behavioral agitation in elderly patients with dementia is problematic both in terms of effectiveness and tolerability.

First, in terms of effect, they do not appear to provide more than minimal benefit in targeting symptoms of agitation, and SGAs may not be different from placebo in this regard.(Yury and Fisher 2007) The NIMH-sponsored CATIE-AD study, which studied the effectiveness of olanzapine, quetiapine and risperidone in the treatment of symptoms of psychosis, aggression and agitation in patients with Alzheimer's disease, also found that even when these symptoms did improve with treatment, the antipsychotic did not improve overall functioning.(Sultzer, Davis, et al. 2008) Furthermore, any improvement in specific symptoms was offset by adverse effects and led to overall discontinuation rates (when both efficacy and tolerability were considered) that were no different from placebo. (Schneider, Tariot, et al. 2006) Secondly, antipsychotics have been found to be associated with an increased risk of stroke in patients with dementia and an overall increased risk of adverse medical events and death in this population. (Gill, Rochon, et al. 2005; Herrmann and Lanctot 2005; Rochon, Normand, et al. 2008; Schneider, Dagerman, et al. 2005) Both FGAs and SGAs appear to increase the risk of death in patients with dementia.(Schneeweiss, Setoguchi, et al. 2007, Wang, Schneeweiss, et al. 2005)

High-potency FGAs are also often used in the treatment of delirium in hospitalized patients.(Lonergan, Britton, et al. 2007) Medical students and physicians should be aware that although antipsychotics may treat the secondary manifestations of delirium, such as behavioral agitation and/or hallucinations, they do not treat the underlying condition. Delirium is a medical condition that is treated by addressing the underlying medical cause.

Further Notes on the Clinical Use of Antipsychotics

All antipsychotics are indicated for the treatment of schizophrenia and are considered reasonably safe and effective for this debilitating disorder. However, there has been much debate about whether there are efficacy differences among these medications, or whether the side effect differences, which are considerable, should be the primary basis for selecting a medication for a particular patient. A meta-analysis of 78 head-to-head comparisons in the literature through 2007 concluded that the efficacy differences are small, but there was some superiority to olanzapine and risperidone, when compared with aripiprazole, quetiapine, and ziprasidone. A problem with this meta-analysis, however, was that almost all of the studies were industry-sponsored. Such studies invariably find outcomes in favor of the sponsor's product, and olanzapine and risperidone have sponsored the largest number of studies.

Another meta-analysis focused on 150 studies that directly compared FGAs with SGAs.(Leucht, Corves, et al. 2009) The authors found that clozapine was clearly superior to the others especially for positive symptoms of hallucinations and delusions. Olanzapine and risperidone were superior to the others, but with a small effect size. The others did not differ in efficacy. The side effect profiles differed markedly, with no pattern to the differences. The authors recommended abandoning the terms "FGA" and "SGA" as irrelevant to efficacy or side effects.

Many clinicians put more reliance on the relatively few comparison studies that were independently funded, such as the CATIE (Clinical Antipsychotics Trials of Intervention Effectiveness) and TEOSS (Treatment of Early-Onset Schizophrenia Spectrum Disorders) studies. (Lieberman, Stroup, et al. 2005; Sikich, Frazier, et al. 2008) Both were funded by the U.S. National Institute of Mental Health. These studies prospectively compared FGAs (perphenazine, molindone) and SGAs (clozapine, olanzapine, risperidone, quetiapine, ziprasidone) and found generally no differences in effectiveness except that clozapine was superior. There were no differences in the ability to improve impaired cognition, despite prior claims for SGA superiority from studies sponsored by the SGA pharmaceutical firms.(Keefe, Bilder, et al. 2007)

Some experts have interpreted CATIE as showing olanzapine to be superior to the other non-clozapine antipsychotics, but this seems likely to be due to peculiar results with the cohort of patients who were on olanzapine prior to entering the CATIE study. These patients (22% of the sample) were randomly assigned to either continue on olanzapine or be switched to one of the other options in CATIE (perphenazine, risperidone, quetiapine, or ziprasidone). The patients who were assigned to remain on olanzapine did better than those who were abruptly switched to any of the other options.(Essock, Covell, et al. 2006) By contrast, the patients who entered the study on risperidone (the second largest group with 19%) showed no advantage to staying on risperidone compared to switching to another agent. Notably, there was no advantage to switching to olanzapine. Hence, the superiority of olanzapine seen in CATIE may be due to the study having a large sample of patients who had been previously stabilized on olanzapine and who clearly responded only to olanzapine or who may have been more prone to a withdrawal-induced exacerbation when taken off of olanzapine. Since olanzapine has a very unfavorable side effect profile with its tendency to promote weight gain, insulin resistance, and the metabolic syndrome, this would appear to make it undesirable as a first-line choice even if it does have slightly superior efficacy.

Antipsychotics are also being used in a wide variety of mood and anxiety disorders as augmentations when antidepressants produce unsatisfactory results, and sometimes they are used as primary treatments for these disorders. Recent data indicating that SGAs and FGAs are associated with double to triple the rate of death from sudden cardiac arrest (presumably from electrophysiological effects related to QT prolongation) suggest that these agents should not be first-line treatments in these clinical situations.(Ray, Chung, et al. 2009) However, antipsychotics are powerful and important options in the treatment of schizophrenia and severe bipolar disorders and these new cardiac concerns should not deter clinicians for prescribing them appropriately for these patients. Obtaining a baseline ECG, and if abnormal obtaining another after dosage has been optimized, is a prudent risk-management approach given this new data.

Table 3 summarizes the characteristics of commonly used antipsychotics.(WHO 2007; PDR 2008; Hyman, Arana, et al. 1995; Rosenbaum, Arana, et al. 2005; Stahl 2005; Taylor, Paton, et al. 2007)

TABLE 3. COMMONLY USED ANTIPSYCHOTICS

MEDICATION*	DOSING**	COMMENTS/ *FDA Indication*
Chlorpromazine (FGA) (Thorazine®)	For oral: Start: 25-50 mg po qhs then increase as tolerated to 300 mg po qhs or in divided doses. Potency: 100 mg po equals haloperidol 2 mg po	Tricyclic structure therefore with TCA side effects, plus EPS; now rarely used as primary antipsychotic; avoid IM given risk of severe orthostasis. On WHO Essential Medicines List for psychotic disorders. *Psychotic disorders/ Other indications (see package insert)*

Thioridazine (FGA) (Mellaril®)	Start: 25 mg po q day/bid/tid for agitation in a variety of anxiety, mood, personality disorders; for schizophrenia increase the same as chlorpromazine. Potency: 80-100 mg po equals haloperidol 2 mg po	Was once the most frequently prescribed antipsychotic; now should avoid use due to one of the highest risks of QTc prolongation of all FGAs and SGAs; doses over 800 mg/day may cause pigmentary retinopathy; CYP2D6 substrate, avoid combining with CYP2D6 inhibitors or any SSRI or propranolol. *Schizophrenia in patients not responsive to or intolerant to other antipsychotics*
Perphenazine (FGA) (Trilafon®)	Start: 4 mg po bid then increase by 4-8 mg every 2 days; 20-24 mg/day in divided doses may be sufficient, 40 mg/day in treatment resistant patients, maximum dose 64 mg/day. Potency: 8-10 mg po equals haloperidol 2mg po	Effective in recent studies in comparison with SGAs; good choice for a first-line FGA. *Schizophrenia*

Molindone (FGA) (Moban®)	Start: 5 mg po bid and increase over several days to typical dose of 30-60 mg total per day, maximum dose 300 mg/day. Potency: 10 mg po equals haloperidol 2 mg po	Recently found to work well in study of adolescents in comparison with risperidone and olanzapine; no weight gain; akathisia is common. *Schizophrenia*
Pimozide (FGA) (Orap®)	Start: 0.5 mg po q day, increase very gradually if needed and maintain low doses (less than stated maximum of 10 mg/day). Potency: 1 mg po equals haloperidol 2 mg po	Avoid use; historically used for delusional parasitosis but no reason to believe better for this than others; high risk of QTc prolongation; CYP3A4 substrate. *Suppression of refractory tics secondary to Tourette's Syndrome*

Fluphenazine (FGA) (Prolixin®)	For oral: Start: 0.5-2 mg po bid and increase as tolerated and necessary, usual daily dose is 5-10 mg/day. PO max is 40 mg/day IM max is 20 mg/day Oral dose is equipotent with haloperidol	Available in short-acting IM for behavioral control and long-acting injectable depot preparation for maintenance treatment of poorly adherent patients given every 2 weeks (see package insert); On WHO Essential Medicines List for psychotic disorders. *Psychotic disorders*
Haloperidol (FGA) (Haldol®)	For oral: Start: 0.5-2 mg po q day or bid and increase as tolerated and necessary, lower doses for elderly delirious patients and higher doses in patients with schizophrenia, 4-10 mg/day may be sufficient in schizophrenia	Most widely used FGA; also used for secondary symptoms of delirium and behavioral control; available in short-acting IM form for behavioral control and long-acting injectable depot form for maintenance treatment given every 4 weeks (see package insert); CYP2D6 substrate. On WHO Essential Medicines List for psychotic disorders. *Psychotic disorders*

| Risperidone (SGA) (Risperdal®, Risperdal M-Tab®, Risperdal Consta®) | For oral risperidone, Risperdal, Risperdal M-Tab: Start: 0.5-1 mg po bid and increase gradually every 1-2 days to target of 4 mg/day, if no response in 1-2 weeks then increase to 6 mg/day, | Fairly well-tolerated SGA, no significant EPS under 4 mg/day in most patients, and medium to low risk of metabolic changes in adults; orthostasis may be a problem initially; hyperprolactinemia is common; may have more rapid action than other SGAs; available in long-acting injectable depot form for maintenance treatment given every 2 weeks (see package insert); CYP2D6 substrate.

Schizophrenia/ Psychotic disorders/ Acute mania or mixed episodes/Irritability from autism (see package insert) |

Olanzapine (SGA) (Zyprexa®, Zydis®, Zyprexa IntraMuscular®)	For oral olanzapine, Zyprexa, Zydis: Start: 15 mg/day for most rapid effect in male smokers for schizophrenia; 10 mg in women smokers; 5 mg in non-smoking women. May increase by 5 mg/day until 15-20 mg/day, (package insert max is 20 mg/day although rarely doses up to 40 mg/day may be slightly better for treatment-resistant cases),see package insert for intramuscular use	Along with clozapine the highest risk of weight gain and metabolic syndrome among SGAs; CYP1A2, CYP2D6 substrate. *Schizophrenia/Acute mania/Bipolar Maintenance*
Quetiapine (SGA) (Seroquel®, Seroquel XR®)	Start: 25-50 mg po bid and double daily until 100 mg bid then increase by 200 mg/day as tolerated depending on sedation and orthostasis to 600-800 mg/day, XR is once-daily version: 200 mg po qhs on day one, 400 mg po qhs on day 2, 600 mg po qhs on day 3	Efficacious in bipolar depression; used frequently off-label as anti-anxiety agent in substance abusers, and in personality disordered; CYP3A4, and CYP2D6 substrate. *Schizophrenia/Acute mania/Bipolar depression/Maintenance treatment of bipolar I disorder as adjunct to lithium or divalproex*

Ziprasidone (SGA) (Geodon®, Geodon for Injection®)	For oral: Start: 20-40 mg po bid and increase dose every 1-2 days to 80 mg po bid (need to take with food for adequate absorption—see text)	SGA with lowest risk of weight gain and metabolic side effects; has higher risk of QTc prolongation; available in short-acting IM form for behavioral control (see package insert); CYP3A4 substrate. *Schizophrenia/Acute mania or mixed episodes*
Aripiprazole (SGA) (Abilify®, Abilify Discmelt®)	For oral: Start: 2.5-5 mg po q am and increase every 2-3 days as tolerated to 15 mg/day, maximum is 30 mg/day but 15 mg/day may be more effective in acute schizophrenia; in mania 15 mg and 30 mg appear equally effective	SGA with low risk of cardiac and metabolic effects; however akathisia is common; very long half-life; available in short-acting IM form for behavioral control (see package insert); CYP2D6 and CYP3A4 substrate. *Schizophrenia/Acute and maintenance treatment of mania and mixed episodes/ Adjunctive therapy to antidepressants for acute treatment of MDD*

Clozapine (SGA) (Clozaril®, Fazaclo®)	Start: 12.5 mg po once or twice daily then increase by 25 mg/day in divided doses as tolerated to 200-400 mg/day and check for response (check serum level if no response—therapeutic serum level of clozapine is over 350 ng/mL, some studies suggest over 450 ng/mL), need to restart at 12.5 mg/day if discontinued 48 hours or more	Risk of agranulocytosis; need WBC/ANC count and ECG before treatment; needs ongoing WBC monitoring—see package insert for WBC monitoring guidelines; multiple other risks; along with lithium may be one of only two drugs with antisuicidal effects; use caution when using with benzodiazepines; do not combine with carbamazepine; *CYP1A2, CYP2D6, CYP3A4 substrate.* *Treatment resistant severe schizophrenia/ Reduction of recurrent suicidal behavior in patients with schizophrenia or schizoaffective disorder*

*Generic and U.S. brand name(s). **Dosing should be adjusted downwards ('start low, go slow' strategy) for the elderly and/or the medically compromised. Abbreviations: ANC-Absolute Neutrophil Count; bid-(bis in die) twice a day; CYP-Cytochrome P450 enzyme; ECG-Electrocardiogram; EPS-Extra-pyramidal Symptoms; FGA-First Generation Antipsychotics; IM-intramuscular; MDD-Major Depressive Disorder; mg-milligram; ng/mL-nanogram per milliliter; po-(per os) orally; q-(quaque) every; qhs-(quaque hora somni) at bedtime; SGA-Second Generation Antipsychotics; TCA-Tricyclic Antidepressants; WBC-White Blood Cell; WHO-World Health Organization.

MOOD STABILIZERS

Lithium and Other Mood Stabilizers

What is a mood stabilizer? Although there is no generally accepted definition, a mood stabilizer can be defined as a medication that can treat either phase of bipolar disorder while not inducing or worsening the other phase. More conservatively, however, a mood stabilizer can be defined as an agent that can both treat *and prevent* both manic and depressive episodes. By this definition only lithium qualifies as a true mood stabilizer. (Bauer and Mitchner 2004)

Lithium (as a salt) has been used as a homeopathic treatment for gout and other disorders since the 1800's. Its calming effect on animals, and subsequently on manic patients, was first described in the 1940s.(Cade 1949) In the brain, lithium inhibits inositol phosphatases that dephosphorylate inositol phosphates that are generated· by the stimulation of G proteins in neuronal membranes activated by a neurotransmitter. This inhibition interferes with inositol regeneration and leads to its depletion in neurons, ultimately leading to decreased neuronal activity.(Nestler, Hyman, et al. 2009) Lithium also inhibits protein kinases, glycogen synthase kinase-3beta, and adenylyl cyclase,(Bachmann, Schloesser, et al. 2005; Lenox and Hahn 2000) and may increase the uptake of the excitatory neurotransmitter glutamate thereby reducing glutamate activity at the neuronal synapse. (Shaldubina, Agam, et al. 2001) Lithium also appears to have neuroprotective properties and may promote neurogenesis.(Chuang 2005; Chen and Manji 2006; Bearden, Thompson, et al. 2007; Nunes, Forlenza, et al. 2007; Fornai, Longone, et al. 2008)

Lithium is effective in both manic and depressive episodes associated with bipolar disorder, as well as for long-term maintenance. It is also the only mood stabilizer with anti-suicidal effect.(Baldessarini, Tondo, et al. 1999; Cipriani, Pretty, et al. 2005) Lithium works particularly well in patients who have a strong family history of bipolar disorder.(Alda 1999)

A target therapeutic serum level of 0.6-0.75mEq/L is recommended for the treatment of bipolar depression and prophylaxis against depressive relapses.(Kleindienst, Severus, et al. 2007; Kleindienst, Severus, et al. 2005; Severus, Kleindienst, et al. 2008) Serum levels of 0.75-1.2mEQ/L may be more effective for the treatment of mania. Serum levels higher than 1.2mEq/L are associated with significant lithium toxicity.

Lithium side effects usually increase with higher serum doses but can occur at any dose. These may include nausea, vomiting, diarrhea, tremor, thirst, polyuria, acne, weight gain, and a benign leukocytosis. Over the long run, lithium can cause hypothyroidism in up to 20% of patients(Johnston and Eagles 1999) (which can be treated with thyroid hormone replacement), and worsening renal function in 20% of patients(Lepkifker, Sverdlik, et al. 2004) (which usually necessitates lithium discontinuation). Because of the many complexities of lithium use, access to references such as the Lithium Encyclopedia for Clinical Practice is recommended. (Jefferson, Greist, et al. 1987)

Physicians should be aware that the anti-manic effects of lithium (and other mood stabilizers, such as valproate and carbamazepine, discussed below) may not be achieved until 7-10 days after a therapeutic dose has been established. In the interim, sedative medications such as antipsychotics and benzodiazepines may be needed when the patient is acutely manic. Once the patient is stabilized, these adjunctive medications can often be tapered and lithium continued as monotherapy.

Valproate (along with carbamazepine and lamotrigine discussed below) is an anticonvulsant with mood stabilizing properties. It is postulated that it exerts its effect via enhancement of GABA transmission. (Johannessen 2000) Despite decades of clinical experience with lithium, valproate has become the most widely used mood stabilizer in the United States. This is primarily due to its ease of use and effectiveness in the treatment of mania. Valproate is not very effective in the treatment of bipolar depression, but has efficacy in the treatment of manic and mixed episodes(Bowden, Brugger, et al. 1994; Freeman, Clothier, et al. 1992) Although serum levels of 50-125 mcg/mL are generally considered to be within the therapeutic range (a range based on anticonvulsant data), the best results in acute mania may occur with levels of greater than 90 mcg/mL. (Allen, Hirschfeld, et al. 2006) Side effects in adults include liver enzyme elevations (usually benign and transient, but not always), weight gain, and possible thrombocytopenia and platelet dysfunction. Because of the latter, bleeding time should be measured prior to surgery even if the platelet count is normal.(De Berardis, Campanella, et al. 2003; Gerstner, Teich, et al. 2006) Valproate is highly protein-bound: concurrent use with warfarin can displace and increase the free fraction of warfarin and increase prothrombin time.

The use of valproate is problematic in women of child bearing age. It is associated with a high risk of teratogenic effects (i.e. neural tube defects).(Cohen 2007; Viguera, Koukopoulos, et al. 2007) Valproate may also play a role in the development of polycystic ovary syndrome. (Joffe, Cohen, et al. 2006; O'Donovan, Kusumakar, et al. 2002)

Many clinicians recommend the use of alternative medications, such as lithium in young bipolar women. Lithium can cause Ebstein's anomaly, an anomaly in the fetal tricuspid valve, at up to 3 times the baseline risk. Previously, this risk was thought to be much higher. For a more complete discussion of pregnancy risks of psychiatric medications, see the 2008 Practice Bulletin of the American College of Obstetrics and Gynecology. (ACOG 2008)

Carbamazepine is an anticonvulsant that can enhance Na+ channel inactivation, thereby blocking action potentials and repetitive neuronal firing. It is also thought to inhibit a process known as kindling—a process whereby repeated subthreshold electrical stimuli can lead to the development of spontaneous seizures. Hypothetically, subthreshold environmental stimuli or prior manias can similarly kindle the development and frequency of further manias.(Rosenbaum, Arana, et al. 2005)

Carbamazepine has efficacy in the treatment of manic,(Weisler, Kalali, et al. 2004; Weisler, Keck, et al. 2005) but probably not depressive,(Ansari and Osser 2009) episodes associated with bipolar disorder. Serum levels of 4-12 mcg/mL may be therapeutic. Side effects such as dizziness, ataxia and gastrointestinal symptoms prohibit the use of loading strategies. Thrombocytopenia, leukopenia, hyponatremia, and dangerous rash may also develop with carbamazepine therapy. Another factor that significantly limits treatment with carbamazepine, especially in severe mania when concurrent antipsychotics may be necessary, is its propensity to induce multiple hepatic enzymes (e.g. CYP1A2, CYP2C9, CYP2C19, CYP3A4). It can therefore increase the metabolism of other concurrently administered drugs and render them less effective. (Notably, the antiepileptic drugs phenobarbital, **phenytoin**, and **primidone** also have similarly broad hepatic enzyme induction capacities.) (Perucca 2006) Teratogenic effects of carbamazepine are comparable in severity to those of valproate.(Cohen 2007; Viguera, Koukopoulos, et al. 2007)

Oxcarbazepine, an analog of carbamazepine, may also have efficacy in the treatment of acute mania.(Ghaemi, Berv, et al. 2003; Pratoomsri, Yatham, et al. 2006) Serum levels do not need to be checked, and there is less enzyme induction with oxcarbazepine, thereby reducing the risk of drug-drug interactions. Hyponatremia, however, remains a concern.(Ortenzi, Paggi, et al. 2008)

Lamotrigine is an anticonvulsant that may inhibit the release of the excitatory amino acid glutamate, but its mechanism of action is not fully known. In bipolar disorder it is often used (but not FDA approved) for the treatment of acute bipolar depression. However, 4 out of 5 studies failed to show separation from placebo. (Calabrese, Bowden, et al. 1999; Calabrese, Huffman, et al. 2008) Nevertheless, it appears to be effective as maintenance therapy for depressive episodes in bipolar disorder.(Bowden, Calabrese, et al. 2003; Calabrese, Bowden, et al. 2003) Although lamotrigine is generally well-tolerated, there is a 0.1% risk of dangerous rash (i.e. toxic epidermal necrolysis—Stevens-Johnson syndrome). Gradual titration is required to decrease the risk of rash. If rash develops, lamotrigine should be discontinued. Lamotrigine so far seems relatively safe in pregnancy. (ACOG 2008)

Second Generation Antipsychotics Used as Mood Stabilizers

All SGAs have been found to have efficacy in the treatment of mania.(Janicak 2006) However, among the SGAs, only quetiapine has been shown to have clear efficacy in the treatment of bipolar depression. (Calabrese, Keck, et al. 2005; Thase, Macfadden, et al. 2006) Olanzapine and aripiprazole have been found effective, as monotherapy treatments, for prevention of manic episodes, and quetiapine has unpublished data suggesting it may prevent both manic and depressive episodes. It may soon present a challenge to lithium as the mood stabilizer with the best evidence base. This remains to be seen.

There is evidence that the SGAs olanzapine, quetiapine, ziprasidone, and aripiprazole, when added to lithium or valproate, can increase maintenance efficacy for the manic phase. In the case of quetiapine, the depressive phase was also helped.

Newer Anticonvulsants

Newer anticonvulsants such as **topiramate**, **gabapentin**, **pregabalin**, **tiagabine**, **zonisamide** and **levetiracetam**, which are generally thought to exert their therapeutic effects by enhancing GABA transmission, may also be effective for the treatment of bipolar disorder.(Johannessen and Landmark 2008) However, when used, they should be considered to be adjunctive treatments only (for example to decrease concurrent anxiety); the evidence base is insufficient to recommend their use as primary agents for the treatment of mood disorder symptoms.(Anand, Bukhari, et al. 2005; Grunze, Langosch, et al. 2003; Grunze, Normann, et al. 2001; Keck, Strawn, et al. 2006; Macdonald and Young 2002; Pande, Crockatt, et al. 2000; Vieta, Goikolea, et al. 2003; Vieta, Manuel Goikolea, et al. 2006; Vieta, Sanchez-Moreno, et al. 2003; Yatham, Kusumakar, et al. 2002; Young, Geddes, et al. 2006; Young, Geddes, et al. 2006)

Arash Ansari, M.D. and David N. Osser, M.D.

Further Notes on the Clinical Approach to Bipolar Patients

Mood stabilization is frequently difficult to achieve in bipolar disorder. Although the goal is to use as few medications as possible and rely on mood stabilizers whenever possible, it is common that more complex psychopharmacology regimens are required. The American Psychiatric Association's recently updated Practice Guidelines for the Treatment of Bipolar Disorder contains a comprehensive review of the current knowledge base on the psychopharmacology of this disorder.(APA 2009)

The Systematic Treatment Enhancement Program – Bipolar Disorder (STEP-BD) is a publically funded project designed to add to our understanding of how to best treat this disorder. The program enrolled 4,360 bipolar patients who are being followed longitudinally at 15 sites. Some of these patients agree to enter controlled studies of a variety of psychosocial and psychopharmacological interventions. Among the significant findings to date are the following:

Psychotherapy is effective for bipolar depression but it is a slow process. It takes 169 days vs. 279 days in the control group.(Miklowitz, Otto, et al. 2007)

Antidepressants (bupropion, paroxetine) are not more effective than placebo for bipolar depression (24% for the antidepressants vs. 27% for the placebo in a 6-month trial). The antidepressants did not induce more switches to mania (10% vs. 11%), but the patients who participated in this study were probably at very low risk for switching.(Sachs, Nierenberg, et al. 2007)

Other STEP-BD data did show that use of antidepressants was associated with more manic symptoms.(Goldberg, Perlis, et al. 2007)

262 suicide attempts and 8 completed suicides have occurred in this patient sample over a 6-year period. Lithium seemed to offer no protective effect, contrary to data from other studies strongly suggesting that lithium helps lower suicide risk in bipolar patients. However, the patient sample clearly had a very low risk of suicidal behaviors so it was not the best population to demonstrate lithium's possible benefit on this symptom.(Marangell, Dennehy, et al. 2008)

Table 4 summarizes the characteristics of commonly used mood stabilizing medications.(Hyman, Arana, et al. 1995; Rosenbaum, Arana, et al. 2005; WHO 2007; PDR 2008; Stahl 2005; Taylor, Paton, et al. 2007)

TABLE 4. COMMONLY USED MOOD STABILIZING MEDICATIONS

MEDICATION*	DOSING**	COMMENTS/ *FDA Indication*
Lithium Carbonate (Lithobid®, Eska-lith®)	Start: 300 mg po bid-tid and check serum trough level after 4-5 days (after steady state) then adjust.	Check baseline chemistries, kidney function, thyroid function (TSH), ECG (r/o sinus node dysfunction); once target dose is reached, check level, chemistries, kidney function, TSH, every 3-6 months initially, then every 6-12 months; NSAIDs, thiazide diuretics, ACE inhibitors, met-ronidazole, and tetracyclines can increase lithium level. On WHO Essential Medicines List for bipolar disorders. *Mania/Maintenance in bipolar disorder*

| Divalproex Sodium, Valproic Acid, Valproate (Depakote®, Depakote ER®, Depakene®) | Start: 250 mg po tid and check serum trough level after 4-5 days, then adjust, can use loading dose of 20-30 mg/kg to hasten response | Check baseline LFTs and CBC; once target dose is reached check serum level, LFTs and CBC every 3-6 months initially then yearly; can inhibit the glucuronidation of lamotrigine; can inhibit CYP2C9, CYP2C19; aspirin can increase levels; valproate is highly protein-bound so will increase free warfarin levels. On WHO Essential Medicines List for bipolar disorders and as an anticonvulsant. *Mania/Mixed episodes associated with bipolar disorder/ Migraine prophylaxis/Specific seizure disorders (see package insert)* |

Carbamazepine (Tegretol®, Carbatrol®, Equetro®)	Start: 200 mg po bid then check serum trough level after 4-5 days. Dose requirements gradually increase over the first month due to cytochrome enzyme induction.	Check baseline CBC, sodium, LFTs; once target dose is reached check serum level, CBC and LFTs every 3-6 months initially, then yearly; induces CYP1A2, CYP2C9, CYP2C19, CYP3A4; itself is a CYP3A4 substrate. On WHO Essential Medicines List for bipolar disorders and as an anticonvulsant. *Acute mania and mixed episodes / Trigeminal neuralgia/Specific seizure disorders (see package insert)*

Lamotrigine (Lamictal®)	Start: 25 mg po q am for first 2 weeks, then 50 mg po q am for 3rd and 4th week, then 100 mg po q am on 5th week, 200 mg po q am on 6th and 7th week, slower titration with concomitant valproate and faster titration and higher doses with concomitant carbamazepine	No laboratory monitoring necessary; valproate and sertraline can increase levels; monitor for rash and Stevens-Johnson Syndrome. *Maintenance treatment for bipolar I disorder in patients treated for acute mood episodes with standard therapy/ Specific seizure disorders (see package insert)*

*Generic and U.S. brand name(s).   **Dosing should be adjusted downwards ('start low, go slow' strategy) for the elderly and/or the medically compromised. Abbreviations: ACE-Angiotensin Converter Enzyme; bid-(bis in die) twice a day; CBC-Complete Blood Count; CYP-Cytochrome P450 enzyme; ECG-Electrocardiogram; kg-kilogram; LFT-Liver Function Tests; mg-milligram; NSAIDS-Non-steroidal Anti-inflammatory Drugs; tid-(ter in die) three times a day; TSH-Thyroid Stimulating Hormone; po-(per os) orally; WHO-World Health Organization.

STIMULANTS AND OTHER ADHD MEDICINES

The use of psychotropics for the treatment of attention-deficit/hyperactivity disorder (ADHD) in children and adolescents is beyond the scope of this chapter. In adults, the diagnosis of ADHD is controversial but guidelines have been developed.(Gibbins and Weiss 2007) Diagnosis of adult ADHD is predicated on plausible historical evidence of childhood onset.(APA, DSM 2000) This is often difficult to establish retrospectively, and when earlier ADHD symptoms are suspected, it is difficult to rule out other etiologies for these symptoms (e.g. family stressors, childhood depression, learning disorders, etc.). Nevertheless, there are adults with undiagnosed ADHD, many of whom have other comorbid psychiatric illnesses, who continue to suffer chronic symptoms through adulthood and may benefit from treatment. Others may have had a clear history and diagnosis of ADHD in childhood and as adults may need to have pharmacological treatments considered or resumed.

Stimulants

Stimulants are the most effective and the first-line treatment for non-substance-abusing patients with ADHD. Amphetamine-like stimulants are sympathomimetic amines that likely enhance norepinephrine and dopaminergic transmission. They may disrupt the presynaptic storage of these transmitters and enhance their release—in both the ascending reticular activating system as well as in the regulation of 'top-down' cortical-thalamic-striatal circuits.(Nestler, Hyman, et al. 2009) **Methylphenidate, dextroamphetamine** and **amphetamine salts** are examples of psychostimulants used in the treatment of ADHD. Assuming correct diagnosis and adequate dose, stimulants' beneficial effects on attentional symptoms, impulsivity and hyperactivity are immediate and subside with medication clearance. Short half-life formulations need to be administered multiple times during the day, but not near bedtime. Stimulant side effects include decreased appetite, insomnia, and anxiety, necessitating gradual dose titration to improve tolerability. Blood pressure and heart rate can also increase with stimulant administration so patients with cardiac disease may not be good candidates for treatment with stimulants. Possible growth retardation and the development of transient tics, although of concern in children, are not likely to be problematic in adults. Chronic stimulant use (or overdose) can lead to psychosis; in susceptible individuals, increased psychosis can be seen after just one dose.(Curran, Byrappa, et al. 2004) Medication interactions of note include that stimulants should not be combined with MAOIs.

The major concern regarding the use of stimulants, however, is the risk of abuse and dependence. This seems related to the stimulants' ability to increase dopaminergic effects in the reward and reinforcement circuitry in the nucleus accumbens. Euphoria, tolerance, and addictive behaviors may develop in susceptible individuals. In the United States, therefore, amphetamine-like stimulants are highly regulated; they are 'Schedule II drugs'-- which indicates that the Drug Enforcement Administration (DEA) designates them as being in the highest risk category for controlled substances that have an established therapeutic use. The risks of addiction and misuse have led some clinicians to be wary of using stimulants even when treatment with these medications is otherwise medically indicated. However, if the diagnosis of ADHD is accurate, these medications should not be avoided in patients who do not have a history of substance abuse. A clear risk and benefit assessment is necessary. Appropriate monitoring and supervision may decrease the risk of abuse. Recent data suggest no increase in risk of subsequent abuse of stimulants when children and adolescents with ADHD are treated with stimulants.(Biederman, Monuteaux, et al. 2008) Notably, there was no decrease in risk either, i.e. no protective effect from treatment. This is controversial however as ADHD is considered to be a risk factor for substance abuse; other data regarding treatment of ADHD as a way to decrease the risk of substance abuse is mixed.(Wilens 2004; Wilens, Faraone, et al. 2003)

Stimulants have also been historically used in the treatment of anergic medically ill, mildly depressed, often elderly patients. Response can often be noted in a matter of days. There is no evidence however that these medications are effective antidepressants in other patients with major depression.(Satel and Nelson 1989)

Non-Stimulant Medicines for ADHD

Atomoxetine is a norepinephrine and dopamine reuptake inhibitor(Bymaster, Katner, et al. 2002) which has shown efficacy in, and has been primarily marketed for, the treatment of ADHD.(Michelson, Adler, et al. 2003) As might be expected by its mechanism of action, it may also have antidepressant effects but there are no published data to support its use in the treatment of depression. Unlike stimulants which can rapidly improve ADHD symptoms, atomoxetine requires several weeks of treatment before response occurs. Response is generally less robust than with stimulants. Atomoxetine may cause increases in blood pressure, insomnia and possible weight loss. It is not associated with abuse or dependence.

In adults, other agents with noradrenergic and/or dopaminergic effects may be helpful in the treatment of ADHD symptoms, although again response is generally weaker than that expected from stimulants.(Meszaros, Czobor, et al. 2007) These include **bupropion**,(Wilens, Spencer, et al. 2001) **tricyclic antidepressants** (especially the more noradrenergic **desipramine** and **nortriptyline**),(Higgins 1999; Prince, Wilens, et al. 2000; Wilens, Biederman, et al. 1996) **venlafaxine**,(Popper 1997) **modafinil** (a wakefulness-promoting drug with unknown mechanism of action)(Biederman, Swanson, et al. 2006) and **clonidine**.(Connor, Fletcher, et al. 1999) SSRIs and antipsychotics are not effective in the treatment of ADHD.

Table 5 summarizes characteristics of selected ADHD medications.(Hyman, Arana, et al. 1995; Rosenbaum, Arana, et al. 2005; PDR 2008; Stahl 2005; Taylor, Paton, et al. 2007)   Antidepressants used in the treatment of ADHD are listed in Table 1.

TABLE 5. SELECTED ADHD MEDICATIONS

MEDICATION*	DOSING**	COMMENTS/ *FDA Indications*
Methylphenidate (Ritalin®, Ritalin LA®, Ritalin SR®, Concerta®, Daytrana®, Metadate CD®, Metadate ER®, Methylin®) And Dexmethylphenidate (Focalin®, Focalin XR®)	For methylphenidate, Ritalin: Start: 5 mg po bid (morning and afternoon) and increase weekly by 10 mg/day, divide bid or tid with last dose not after 6 pm, maximum 60 mg/day with bid-tid dosing.	Carries risk of abuse; may decrease appetite and cause insomnia. *Treatment of ADHD and narcolepsy*
Amphetamine salts (Adderall®, Adderall XR®)	For Adderall: Start: 5 mg po q am and increase weekly by 5 mg/day, maximum 60 mg/day with bid dosing (morning and afternoon)	Carries risk of abuse; may decrease appetite and cause insomnia *Treatment of ADHD and narcolepsy*

| Dextroamphet-amine (Dexedrine®, Dextrostat®) | Start: 5 mg po q am and increase weekly by 10 mg/day, maximum 40 mg/day with bid dosing (morning and afternoon) | Carries risk of abuse; may decrease appetite and cause insomnia. *Treatment of ADHD and narcolepsy* |
| Atomoxetine (Strattera®) | Start: 40 mg po q am or divided bid (morning and afternoon), after 3 days increase to 80 mg/day, maximum 100 mg/day, reduced dosing with hepatic insufficiency | No risk of abuse; slower response than with stimulants; CYP2D6 substrate. *Treatment of ADHD* |

*Generic and U.S. brand name(s). **Dosing should be adjusted downwards ('start low, go slow' strategy) for the elderly and/or the medically compromised. Abbreviations: ADHD-Attention Deficit/Hyperactivity Disorder; bid-(bis in die) twice a day; CYP-Cytochrome P450 enzyme; mg-milligram; po-(per os) orally; tid-(ter in die) three times a day; q-(quaque) every.

TREATMENTS FOR SUBSTANCE ABUSE/ DEPENDENCE

The past few decades have seen a dramatic increase in the number of pharmacological options available for the treatment of substance abuse disorders. Pharmacotherapeutic treatments are now available for the treatment of opioid dependence, alcohol dependence and nicotine dependence. Detailed algorithms for the use of pharmacotherapy in addiction disorders may be found at the website of the International Psychopharmacology Algorithm Project (www.ipap.org). The medical treatment of withdrawal states that emerge upon substance discontinuation are not covered in this chapter, but have been reviewed elsewhere.(Rosenbaum, Arana, et al. 2005; Taylor, Paton, et al. 2007) In treating patients with substance abuse, the beneficial effects of psychosocial interventions should not be overlooked. (Dutra, Stathopoulou, et al. 2008)

Arash Ansari, M.D. and David N. Osser, M.D.

Medications for Opioid Dependence

Methadone, a synthetic opioid mu-receptor agonist, is a long-acting analgesic that has shown efficacy in maintenance therapy for patients with a history of opioid dependence. Although methadone (at relatively low doses) can be prescribed as an analgesic by individual physicians in the United States, methadone for the treatment of heroin dependence can only be dispensed by centers registered and authorized to do so by regulatory agencies. The methadone dose is gradually increased over many months in patients attending these centers until a dose (of usually 60-120 mg/day or higher) is reached that stops cravings for illicit opiates and stops drug seeking behaviors.(Faggiano, Vigna-Taglianti, et al. 2003) Methadone can cause respiratory depression (especially in patients who are not tolerant to opioids), additive CNS effects with concurrent use of other sedatives, QT prolongation(Ehret, Voide, et al. 2006) and constipation. Levomethadylacetate (**LAAM**) is similar to methadone except with a longer half-life requiring less than daily administration, but it has been discontinued due to reports of it causing torsades de pointes arrhythmias.

Physicians should be aware that patients maintained on high dose methadone who are admitted to the medical/surgical wards of hospitals for unrelated medical care are likely to need to continue their daily dose of methadone. However, high doses should never be administered without independent confirmation with the methadone center administering this drug to confirm the actual dose that the patient has been receiving prior to admission. Even 3-4 days of methadone discontinuation may significantly reduce patients' tolerance to the respiratory effects of this drug. To decrease the risk of death from respiratory depression, a single dose of methadone should never exceed 20 mg when independent confirmation of higher doses is not possible. Subsequent dose increments can then be added as necessary and as tolerated.

Buprenorphine is an opioid mu-receptor partial agonist (with very high affinity for this receptor) that is used as an alternative to methadone for maintenance therapy in opioid dependence.(Fudala, Bridge, et al. 2003) In the United States it can be prescribed in an office-based setting (for example with weekly counseling and weekly dispensing)(Fiellin, Pantalon, et al. 2006) without requiring daily administration in a methadone center. Buprenorphine is less dangerous than methadone in overdose with a lower risk of respiratory depression. Concurrent use of benzodiazepines or alcohol however significantly increases the risk of death from respiratory depression;(Megarbane, Hreiche, et al. 2006; Kintz 2001) therefore patients with a history of polysubstance abuse may not be good candidates for buprenorphine maintenance therapy. When used in outpatient treatment buprenorphine is combined with the opioid antagonist **naloxone** and administered sublingually. In sublingual form the buprenorphine is absorbed while the naloxone is not. When absorbed through the GI tract, naloxone undergoes extensive first-pass liver metabolism decreasing its systemic availability. Naloxone is added to discourage *intravenous* abuse of this medication: if this combination is misused intravenously, the naloxone effect predominates and blocks any opioid effect. Buprenorphine may also be beneficial in chronic pain patients who are at risk of opioid dependence; higher doses may be needed when buprenorphine is used as an analgesic.

Medications for Alcohol Dependence

Disulfiram, one of the earliest treatments developed for addictive disorders, acts by producing unpleasant physical effects if alcohol is concurrently consumed. It disrupts ethanol metabolism by irreversibly inhibiting aldehyde dehydrogenase, thereby leading to a significant accumulation of ethanol metabolite acetaldehyde which is associated with severely unpleasant adverse effects (and cardiac stress). Although there is no evidence that it helps maintain abstinence over the long run, it may be useful as a disincentive to ethanol use in the short term.(Suh, Pettinati, et al. 2006) It retains its effect on aldehyde dehydrogenase for up to 2 weeks, so even if the patient stops taking disulfiram and plans to drink, there may be time to reconsider and enlist other supportive mechanisms to maintain sobriety before it loses effectiveness. Ultimately however, most patients who wish to drink do so by discontinuing disulfiram, and many drink while still on it, placing themselves at severe risk. Therefore, like all pharmacotherapies for ethanol dependence, external supports (such as family supervision of medication adherence) and nonpharmacological therapies (such as ongoing counseling and behavioral therapies) are needed for continued effectiveness.(Lingford-Hughes, Welch, et al. 2004; Hughes and Cook 1997) Notably, however, a recent randomized comparison of disulfiram, acamprosate and naltrexone (discussed below) in 243 patients, all of whom received brief cognitive-behavioral psychotherapy, showed disulfiram to be more advantageous than the other agents.(Laaksonen, Koski-Jannes, et al. 2008)

Patients who are beginning disulfiram treatment should be informed of possible medication interactions and the need for avoidance of alcohol in foods (e.g. sauces) and topical preparations (e.g. perfumes). Disulfiram is not recommended for patients with cardiac disease, significant liver disease, peripheral neuropathy or psychosis.

Acamprosate may increase the number of abstinence days and decrease overall alcohol consumption long-term in alcohol dependent patients.(Boothby and Doering 2005; Sass, Soyka, et al. 1996; Whitworth, Fischer, et al. 1996; Mann, Lehert, et al. 2004; Kranzler and Van Kirk 2001) Its mechanism of action is unclear although it is thought to involve the enhancement of GABA transmission and possibly the antagonism of the excitatory neurotransmitter glutamate.(Littleton and Zieglgansberger 2003) It is generally well-tolerated, with mild GI symptoms as the most commonly seen adverse effects. It is renally excreted and may be administered to patients with liver disease. Recent evidence from a large multicenter study, however, has shed doubt on the effectiveness of acamprosate(Anton, O'Malley, et al. 2006)—see below.

Naltrexone is an opioid receptor antagonist. Alcohol can increase the release of endogenous opioids in the brain which may contribute to its euphoric effects. Naltrexone may reduce this opioid-mediated aspect of alcohol's reinforcing properties, and modestly reduce alcohol use in dependent patients.(Anton 2008; Srisurapanont and Jarusuraisin 2005) It appears to be most beneficial in severe alcoholics.(Pettinati, O'Brien, et al. 2006) A long-acting (i.e. every 4 weeks) injectable preparation is also available.(Garbutt, Kranzler, et al. 2005; O'Malley, Garbutt, et al. 2007) Naltrexone may cause mild GI symptoms and an infrequent transaminitis that requires monitoring. Patients on naltrexone must not be given opiates for pain management: overdose and death can result from the high opiate doses needed to override the effect of naltrexone.

Some studies have suggested superior efficacy of naltrexone as compared to acamprosate.(Anton, O'Malley, et al. 2006; Morley, Teesson, et al. 2006; Rubio, Jimenez-Arriero, et al. 2001) The recent U.S. government-sponsored COMBINE study which compared naltrexone vs. acamprosate vs. the combination of the two, all combined with medical management (i.e. brief meetings with a healthcare provider modeled on a primary care setting), found naltrexone to be more effective than acamprosate. It also found that the meetings with a healthcare provider increased the likelihood of abstinence.(Anton, O'Malley, et al. 2006) It should be noted that the dose of naltrexone used was twice the usual maximum dose (100 mg/day vs. 50 mg/day). Individuals with a specific polymorphism of the mu-opioid receptor gene (OPRM1), i.e. individuals with an Asp40 allele—coding for a receptor with increased beta-endorphin binding and activity(Bond, LaForge, et al. 1998)—may be more likely to respond to naltrexone. (Anton, Oroszi, et al. 2008)

Off-label use of naltrexone may also be considered in opioid dependent patients—where higher doses are usually needed—if there are significant external supports and motivation to ensure adherence to this medication(Kirchmayer, Davoli, et al. 2003). Highly motivated addicted physicians and other professionals are examples of patients who may benefit from naltrexone treatment for opiate dependency.(Ling and Wesson 1984; Washton, Gold, et al. 1984)

Other Medications for Alcohol Dependence

There is recent evidence to support the use of the anticonvulsant **topiramate** for the treatment of alcohol dependence.(Johnson, Ait-Daoud, et al. 2003; Johnson, Rosenthal, et al. 2007) **Ondansetron**, an antiemetic with 5HT3 serotonin receptor antagonist activity, has also emerged as an agent with possible efficacy for this indication.(Johnson, Ait-Daoud, et al. 2000; Johnson, Roache, et al. 2000) The antidepressant mirtazapine is also a 5HT3 antagonist but has not been studied for this indication.

Medications for Nicotine Dependence

Nicotine replacement therapy is used to decrease withdrawal symptoms during smoking tapering and cessation and can double the odds of quitting.(Silagy, Lancaster, et al. 2004) Nicotine replacement can be delivered transdermally via a patch, or by gum, oral inhaler, nasal spray or dissolving lozenge. All modes of delivery are likely to be effective(Silagy, Lancaster, et al. 2004) Actual dosing and duration of treatment vary slightly for each formulation, although all nicotine replacement treatments involve setting a target date for smoking cessation followed by a gradual taper of the nicotine over 2-3 months. In a review of 88 trials, success rates on 6-12 month follow-up averaged 16% vs. 10% on placebo.(Silagy, Mant, et al. 2000) Patients should therefore be encouraged to make repeated efforts to quit. Caution should be used in patients with a history of cardiac disease, especially when using the nicotine patch (avoid the patch if there is a history of serious arrhythmias, angina or immediately post-MI). Patients should not smoke at all while wearing the transdermal nicotine patch. Nausea and headaches can occur frequently with nicotine replacement therapy.

Bupropion, an antidepressant with possible dopaminergic effects (see section on antidepressants), is also efficacious for smoking cessation.(Jorenby, Leischow, et al. 1999) Bupropion should be started two weeks before the target stop date and then continued for at least 3 months. The addition of nicotine replacement therapy to bupropion can increase the chances of abstinence compared to the use of either drug alone.(Jorenby, Leischow, et al. 1999)

Varenicline is an alpha4beta2 nicotinic acetylcholine receptor partial agonist, with high affinity for this receptor. It is the latest advance in nicotine addiction treatment. It may have effectiveness that is comparable to, or greater than that of bupropion for smoking cessation.(Gonzales, Rennard, et al. 2006; Jorenby, Hays, et al. 2006; Tonstad, Tonnesen, et al. 2006) Although varenicline appears to be generally well-tolerated, treatment-emergent mood changes and psychosis have been reported in susceptible patients.(Freedman 2007; Kohen and Kremen 2007) Because of its cost it is often restricted by pharmacy benefit managers to patients who have failed nicotine replacement and bupropion therapy.

Table 6 summarizes the characteristics of medications used for substance abuse/dependence disorders. (WHO 2007; PDR 2008; Hyman, Arana, et al. 1995; Rosenbaum, Arana, et al. 2005; Stahl 2005; Taylor, Paton, et al. 2007)

TABLE 6. MEDICATIONS FOR SUBSTANCE ABUSE/DEPENDENCE

MEDICATION*	DOSING**	COMMENTS/ *FDA Indication*
Methadone (Opioid analgesic) (Dolophine®, Methadose®)	Gradually increased over many months at specialized methadone maintenance centers only, to reach a dose that would stop cravings for illicit opiates—see text; analgesic doses are much lower	The use of prescribed opiates for addicts is controversial, but effective; not curative; requires attendance at a methadone clinic for daily administration; may increase QTc; CYP3A4 substrate. On WHO Essential Medicines List for substance dependence. *Detoxification and temporary maintenance treatment of narcotic addiction/ Relief of severe pain*

Buprenorphine with and without naloxone (Partial opioid agonist with or without opioid antagonist) (Suboxone®, Subutex®)	Do not start until patient is experiencing moderate opiate withdrawal. Start: 4 mg sublingually bid-tid, usual maintenance dose is 16-20 mg/day or less, in divided doses	May be given as take home prescription by trained physicians; less regulated than methadone, and considerable street role and usage is occurring; CYP3A4 substrate. *Opioid dependence*
Acamprosate (GABA analog) (Campral®)	Start: 333 mg po tid and increase to 666 mg po tid after 2-3 days	Renally cleared; check baseline kidney function and adjust dose with decreased function; can continue even with alcohol relapse; concurrent naltrexone may increase serum levels; suicidality rates higher in acamprosate treated patients compared to placebo groups. *Maintenance of abstinence from alcohol in patients with alcohol dependence who are abstinent at treatment initiation*

Disulfiram (Aldehyde Dehydrogenase Inhibitor) (Antabuse®)	Start 24 hours or longer after last alcohol use. Start: 125 mg po q am and increase after 4 days to 250 mg po q am and continue, maximum 500 mg/day	Check baseline LFTs before treatment and after 2 weeks and then every 3-6 months thereafter. *Aid in the management of selected chronic alcoholics who want to remain sober and commit to supportive and psychotherapeutic treatment*
Naltrexone (Opioid antagonist) (ReVia®, Vivitrol®)	For oral naltrexone, ReVia: Start: 25 mg po q am after meal then increase to 50 mg po q am after 3 days, do not start until free from opioids for 7-10 days	Check baseline LFTs, then monthly for 3 months, then every 3-6 months thereafter; available in long-acting IM form for every 4 weeks administration; give patient medi-alert card or bracelet. *Treatment of alcohol dependence and to block effects of exogenously administered opioids*

Bupropion (Antidepressant) (Zyban®, Wellbutrin®, Wellbutrin SR®, Wellbutrin XL®)	For bupropion, Zyban, Wellbutrin: Start while still smoking. Start: 75-100 mg po bid (morning and afternoon) and increase after 4-7 days to 150 mg po bid, set cigarette cessation target date 2 weeks into treatment.	May be combined with nicotine replacement therapy; CYP2D6 inhibitor. *Aid to smoking cessation treatment/ MDD/Prevention of MDE in patients with seasonal affective disorder*
Varenicline (Nicotine Acetylcholine Receptor Agonist) (Chantix®)	Start: 0.5 mg po bid for 7 days, then 1 mg po bid and continue for 12-24 weeks	Treatment-emergent neuropsychiatric symptoms and suicidality reported. *Aid to smoking cessation treatment*

Nicotine (Commit®, Nicoderm®, Nicorette®, Nicotrol Inhaler®, Nicotrol Nasal Spray®)	For Nicoderm patch: Stop smoking, then dosing depends on cigarette use: Ex.: If greater than 10 cigarettes/day then: 21 mg patch TD each day for 6 weeks, then 14 mg TD each day for 2 weeks then 7 mg TD each day for 2 weeks then stop, other dosing depends on formulation	Nicotine replacement therapy also serves to eliminate hydrocarbon toxicity and carbon monoxide inhalation associated with cigarette use. *To reduce withdrawal symptoms associated with smoking cessation*

*Generic and U.S. brand name(s). **Dosing should be adjusted downwards ('start low, go slow' strategy) for the elderly and/or the medically compromised. Abbreviations: bid-(bis in die) twice a day; CYP-Cytochrome P450 enzyme; FDA-Food and Drug Administration; GABA-Gamma-Aminobutyric Acid; IM-intramuscular; LFT-Liver Function Tests; MDD-Major Depressive Disorder; MDE-Major Depressive Episode; mg-milligram; po-(per os) orally; q-(quaque) every; TD-transdermally; tid-(ter in die) three times a day; WHO-World Health Organization.

CONCLUSION

Over the last five decades, multiple medications have become available for the treatment of patients with psychiatric disorders. TCAs, MAOIs, SSRIs, SNRIs and others antidepressants have expanded current treatment options for depressive and anxiety disorders. Anxiolytics, including benzodiazepines and non-dependence-producing alternatives, are available for the treatment of severe anxiety disorders. First and second generation antipsychotics with different receptor profiles and side effect profiles have expanded the choices for patients with psychotic disorders. Lithium and other medications with mood stabilizing properties are available for use in patients with bipolar disorder. New formulations of stimulants and non-stimulant agents can be used in adults with attention-deficit/hyperactivity disorder. Finally, pharmacological therapies for the treatment of substance abuse and dependence disorders have been greatly expanded in recent years.

Medical students and physicians should become familiar with these medications and obtain some facility in using them. As always, the science and art of medicine comprise the ability to appropriately and carefully apply that which is learned in textbooks to a specific patient. In the clinical setting, pharmacotherapeutic treatments should be used judiciously: the risks and benefits of treatments should be considered so that every effort is made to 'first do no harm.' One should employ the strategy of using one medication at a time, so as to have the opportunity to know what is actually working and not working. The goal is to provide relief and lessen suffering, preferably in the most evidence-based and cost-effective manner possible. Finally, students and clinicians should keep in mind that for pharmacotherapeutic interventions to be successful, there must also be appropriate psychosocial support and treatment. Only then can safe, effective, and comprehensive treatment be provided.

BIBLIOGRAPHY

ACOG-American College of Obstetricians and Gynecologists Practice Bulletin: Clinical management guidelines for obstetrician-gynecologists, Number 92, April 2008. Use of psychiatric medications during pregnancy and lactation. Obstetrics and Gynecology 111:1001-1020, 2008

ADA-American Diabetes Association, American Psychiatric Association, American Association of Clinical Endocrinologists, North American Association for the Study of Obesity: Consensus development conference on antipsychotic drugs and obesity and diabetes. Journal of Clinical Psychiatry 65:267-272, 2004

Alda M: Pharmacogenetics of lithium response in bipolar disorder. Journal of Psychiatry and Neuroscience 24:154-158, 1999

Allen MH, Hirschfeld RM, et al.: Linear relationship of valproate serum concentration to response and optimal serum levels for acute mania. American Journal of Psychiatry 163:272-275, 2006

Anand A, Bukhari L, et al.: A preliminary open-label study of zonisamide treatment for bipolar depression in 10 patients. Journal of Clinical Psychiatry 66:195-198, 2005

Ansari A: The efficacy of newer antidepressants in the treatment of chronic pain: a review of current literature. Harvard Review of Psychiatry 7:257-277, 2000

Ansari A, Osser, DN: The Psychopharmacology Algorithm Project at the Harvard South Shore Program: An Update on Bipolar Depression. Submitted for publication, 2009

Ansari A, Osser DN, Lai LS, Schoenfeld PM, Potts KC: Pharmacological approach to the psychiatric inpatient, in Ovsiew F, Munich RL (eds), Principles of Inpatient Psychiatry. Philadelphia, PA: Lippincott Williams & Wilkins, 2009

Anton RF: Naltrexone for the management of alcohol dependence. New England Journal of Medicine 359:715-721, 2008

Anton RF, O'Malley SS, et al.: Combined pharmacotherapies and behavioral interventions for alcohol dependence: the COMBINE study: a randomized controlled trial. Journal of the American Medical Association 295:2003-2017, 2006

Anton RF, Oroszi G, et al.: An evaluation of mu-opioid receptor (OPRM1) as a predictor of naltrexone response in the treatment of alcohol dependence: results from the Combined Pharmacotherapies and Behavioral Interventions for Alcohol Dependence (COMBINE) study. Archives of General Psychiatry 65:135-144, 2008

APA-American Psychiatric Association: American Psychiatric Association Practice Guidelines for the Treatment of Patients with Panic Disorder. American Journal of Psychiatry 155(Suppl 5):1-34, 1998

APA-American Psychiatric Association: (DSM IV) Diagnostic and Statistical Manual of Mental Disorders, Fourth Edition, Text Revision. Washington, D.C.: American Psychiatric Association, 2000

APA-American Psychiatric Association: The American Psychiatric Association Practice Guidelines for the Treatment of Patients with Major Depressive Disorder. American Journal of Psychiatry 157(Suppl 4), 2000

APA-American Psychiatric Association: American Psychiatric Association Practice Guidelines for the Treatment of Patients with Acute Stress Disorder and Posttraumatic Stress Disorder. American Journal of Psychiatry 161:1-31, 2004

APA-American Psychiatric Association: American Psychiatric Association Practice Guidelines for the Treatment of Bipolar Disorder. American Journal of Psychiatry 166, 2009

Bachmann RF, Schloesser RJ, et al.: Mood stabilizers target cellular plasticity and resilience cascades: implications for the development of novel therapeutics. Molecular Neurobiology 32:173-202, 2005

Baldessarini RJ, Tondo L, et al.: Effects of lithium treatment and its discontinuation on suicidal behavior in bipolar manic-depressive disorders. Journal of Clinical Psychiatry 60 (Suppl 2):77-84, 1999

Banerjee S, Shamash K, et al.: Randomized controlled trial of effect of intervention by psychogeriatric team on depression in frail elderly people at home. British Medical Journal 313:1058-61, 1996

Baptista T, Rangel N, et al.: Metformin as an adjunctive treatment to control body weight and metabolic dysfunction during olanzapine administration: a multicentric, double-blind, placebo-controlled trial. Schizophrenia Research 93:99-108, 2007

Barkham M, Hardy GE: Counseling and interpersonal therapies for depression: towards securing an evidence-base. British Medical Bulletin 57:115-132, 2001

Bauer MS, Mitchner L: What is a "mood stabilizer"? An evidence-based response. American Journal of Psychiatry 161:3-18, 2004

Bearden CE, Thompson PM, et al.: Greater cortical gray matter density in lithium-treated patients with bipolar disorder. Biological Psychiatry 62:7-16, 2007

Bertilsson L: Geographical/interracial differences in polymorphic drug oxidation. Current state of knowledge of cytochromes P450 (CYP) 2D6 and 2C19. Clinical Pharmacokinetics 29:192-209, 1995

Biederman J, Monuteaux MC, et al.: Stimulant therapy and risk for subsequent substance use disorders in male adults with ADHD: a naturalistic controlled 10-year follow-up study. American Journal of Psychiatry 165:597-603, 2008

Biederman J, Swanson JM, et al.: A comparison of once-daily and divided doses of modafinil in children with attention-deficit/hyperactivity disorder: a randomized, double-blind, and placebo-controlled study. Journal of Clinical Psychiatry 67:727-735, 2006

Boehnlein JK, Kinzie JD: Pharmacologic reduction of CNS noradrenergic activity in PTSD: the case for clonidine and prazosin. Journal of Psychiatric Practice 13:72-78, 2007

Bond C, LaForge KS, et al.: Single-nucleotide polymorphism in the human mu opioid receptor gene alters beta-endorphin binding and activity: possible implications for opiate addiction. Proceedings of the National Academy of Sciences of the United States of America 95:9608-13, 1998

Boothby LA, Doering PL: Acamprosate for the treatment of alcohol dependence. Clinical Therapeutics 27:695-714, 2005

Bowden CL, Brugger AM, et al.: Efficacy of divalproex vs. lithium and placebo in the treatment of mania. The Depakote Mania Study Group. Journal of the American Medical Association 271:918-924, 1994

Bowden CL, Calabrese JR, et al.: A placebo-controlled 18-month trial of lamotrigine and lithium maintenance treatment in recently manic or hypomanic patients with bipolar I disorder. Archives of General Psychiatry 60:392-400, 2003

Boyer EW, Shannon M: The serotonin syndrome. New England Journal of Medicine 352:1112-1120, 2005

Bridge JA, Iyengar S, et al.: Clinical response and risk for reported suicidal ideation and suicide attempts in pediatric antidepressant treatment: a meta-analysis of randomized controlled trials. Journal of the American Medical Association 297:1683-1696, 2007

Buffett-Jerrott SE, Stewart SH: Cognitive and sedative effects of benzodiazepine use. Current Pharmaceutical Design 8:45-58, 2002

Bymaster FP, Katner JS, et al.: Atomoxetine increases extracellular levels of norepinephrine and dopamine in prefrontal cortex of rat: a potential mechanism for efficacy in attention deficit/hyperactivity disorder. Neuropsychopharmacology 27:699-711, 2002

Cade JF: Lithium salts in the treatment of psychotic excitement. Medical Journal of Australia 2:349-352, 1949

Calabrese JR, Bowden CL, et al.: A placebo-controlled 18-month trial of lamotrigine and lithium maintenance treatment in recently depressed patients with bipolar I disorder. Journal of Clinical Psychiatry 64:1013-1024, 2003

Calabrese JR, Bowden CL, et al.: A double-blind placebo-controlled study of lamotrigine monotherapy in outpatients with bipolar I depression. Lamictal 602 Study Group. Journal of Clinical Psychiatry 60:79-88, 1999

Calabrese JR, Huffman RF, et al.: Lamotrigine in the acute treatment of bipolar depression: results of five double-blind, placebo-controlled clinical trials. Bipolar Disorders 10:323-333, 2008

Calabrese JR, Keck PE, et al.: A randomized, double-blind, placebo-controlled trial of quetiapine in the treatment of bipolar I or II depression. American Journal of Psychiatry 162:1351-1360, 2005

Chaudron LH, Pies RW: The relationship between postpartum psychosis and bipolar disorder: a review. Journal of Clinical Psychiatry 64:1284-1292, 2003

Chen G, Manji HK: The extracellular signal-regulated kinase pathway: an emerging promising target for mood stabilizers. Current Opinion in Psychiatry 19:313-323, 2006

Chiu CC, Chen KP, et al.: The early effect of olanzapine and risperidone on insulin secretion in atypical-naïve schizophrenic patients. Journal of Clinical Psychopharmacology 26:504-507, 2006

Chuang DM: The antiapoptotic actions of mood stabilizers: molecular mechanisms and therapeutic potentials. Annals of the New York Academy of Sciences 1053:195-204, 2005

Cipriani A, Furukawa TA, et al.: Comparative efficacy and acceptability of 12 new-generation antidepressants: a multiple-treatments meta-analysis. www.thelancet.com published online January 29, 2009

Cipriani A, Pretty H, et al.: Lithium in the prevention of suicidal behavior and all-cause mortality in patients with mood disorders: a systematic review of randomized trials. American Journal of Psychiatry 162:1805-1819, 2005

Cohen LS: Treatment of bipolar disorder during pregnancy. Journal of Clinical Psychiatry 68(Suppl 9):4-9, 2007

Connor DF, Fletcher KE, et al.: A meta-analysis of clonidine for symptoms of attention-deficit hyperactivity disorder. Journal of the American Academy of Child and Adolescent Psychiatry 38:1551-1559, 1999

Correll CU, Leucht S, et al.: Lower risk of tardive dyskinesia associated with second-generation antipsychotics: a systematic review of 1-year studies. American Journal of Psychiatry 161:414-425, 2004

Cubala WJ, Landowski J: Seizure following sudden zolpidem withdrawal. Progress in Neuropsychopharmacology and Biological Psychiatry 31:539-540, 2007

Cuijpers P, Van Straten A, et al.: Are psychological and pharmacologic interventions equally effective in the treatment of adult depressive disorders? A meta-analysis of comparative studies. Journal of Clinical Psychiatry 69:1675-1685, 2009

Curran C, Byrappa N, et al.: Stimulant psychosis: systematic review. British Journal of Psychiatry 185:196-204, 2004

De Berardis D, Campanella D, et al.: Thrombocytopenia during valproic acid treatment in young patients with new-onset bipolar disorder. Journal of Clinical Psychopharmacology 23:451-458, 2003

DeVane CL, Grothe DR, et al.: Pharmacology of antidepressants: focus on nefazodone. Journal of Clinical Psychiatry 63(Suppl 1):10-17, 2002

Dutra L, Stathopoulou G, et al.: A meta-analytic review of psychosocial interventions for substance use disorders. American Journal of Psychiatry 165:179-187, 2008

Ehret GB, Voide C, et al.: Drug-induced long QT syndrome in injection drug users receiving methadone: high frequency in hospitalized patients and risk factors. Archives of Internal Medicine 166:1280-1287, 2006

El-Sayeh HG, Morganti C, et al.: Aripiprazole for schizophrenia. Systematic review. British Journal of Psychiatry 189:102-108, 2006

Ereshefsky L, Jhee S, et al.: Antidepressant drug-drug interaction profile update. Drugs R & D 6:323-336, 2005

Essock SM, Covell NH, et al.: Effectiveness of switching antipsychotic medications. American Journal of Psychiatry 163:2090-2095, 2006

Faggiano F, Vigna-Taglianti F, et al.: Methadone maintenance at different dosages for opioid dependence. Cochrane Database of Systematic Reviews CD002208, 2003

Farde L, Nordstrom AL, et al.: Positron emission tomographic analysis of central D1 and D2 dopamine receptor occupancy in patients treated with classical neuroleptics and clozapine. Relation to extrapyramidal effects. Archives of General Psychiatry 49:538-544, 1992

Fava M, Rush AJ, et al.: Difference in treatment outcome in outpatients with anxious versus nonanxious depression: a STAR*D report. American Journal of Psychiatry 165:342-351, 2008

Fava M, Rush AJ, et al.: A comparison of mirtazapine and nortriptyline following two consecutive failed medication treatments for depressed outpatients: a STAR*D report. American Journal of Psychiatry 163:1161-1172, 2006

Fayek M, Kingsbury SJ, et al.: Cardiac effects of antipsychotic medications. Psychiatric Services 52:607-609, 2001

Feighner JP: Cardiovascular safety in depressed patients: focus on venlafaxine. Journal of Clinical Psychiatry 56:574-579, 1995

Feighner JP: Mechanism of action of antidepressant medications. Journal of Clinical Psychiatry 60(Suppl 4):4-11, 1999

Ferreri M, Hantouche EG, et al.: Value of hydroxyzine in generalized anxiety disorder: controlled double-blind study versus placebo. L' Encephale 20:785-791, 1994

Fiellin DA, Pantalon MV, et al.: Counseling plus buprenorphine-naloxone maintenance therapy for opioid dependence. New England Journal of Medicine 355:365-374, 2006

Fishbain D: Evidence-based data on pain relief with antidepressants. Annals of Medicine 32:305-316, 2000

Fornai F, Longone P, et al.: Lithium delays progression of amyotrophic lateral sclerosis. Proceedings of the National Academy of Sciences of the United States of America 105: 2052-2057, 2008

Freedman R: Exacerbation of schizophrenia by varenicline. American Journal of Psychiatry 164:1269, 2007

Freeman TW, Clothier JL, et al.: A double-blind comparison of valproate and lithium in the treatment of acute mania. American Journal of Psychiatry 149:108-111, 1992

Fudala PJ, Bridge TP, et al.: Office-based treatment of opiate addiction with a sublingual-tablet formulation of buprenorphine and naloxone. New England Journal of Medicine 349:949-958, 2003

Garbutt JC, Kranzler HR, et al.: Efficacy and tolerability of long-acting injectable naltrexone for alcohol dependence: a randomized controlled trial. Journal of the American Medical Association 293:1617-1625, 2005

Gartlehner G, Gaynes BN, et al.: Comparative benefits and harms of second-generation antidepressants: background paper for the American College of Physicians. Annals of Internal Medicine 149:734-750, 2008

Gelenberg AJ: Nefazodone hepatotoxicity: Black Box Warning. Biological Therapies in Psychiatry Newsletter 25, 2002

Gerstner T, Teich M, et al.: Valproate-associated coagulopathies are frequent and variable in children. Epilepsia 47:1136-1143, 2006

Ghaemi SN, Berv DA, et al.: Oxcarbazepine treatment of bipolar disorder. Journal of Clinical Psychiatry 64:943-945, 2003

Ghaemi SN, Ko JY, et al.: "Cade's disease" and beyond: misdiagnosis, antidepressant use, and a proposed definition for bipolar spectrum disorder. Canadian Journal of Psychiatry 47:125-134, 2002

Gibbins C, Weiss M: Clinical recommendations in current practice guidelines for diagnosis and treatment of ADHD in adults. Current Psychiatry Reports 9:420-426, 2007

Gibbons RD, Brown CH, et al.: Early evidence on the effects of regulators' suicidality warnings on SSRI prescriptions and suicide in children and adolescents. American Journal of Psychiatry 164:1356-1363, 2007

Gill SS, Rochon PA, et al.: Atypical antipsychotic drugs and risk of ischemic stroke: population based retrospective cohort study. British Medical Journal 330:445, 2005

Glass J, Lanctot KL, et al.: Sedative hypnotics in older people with insomnia: meta-analysis of risks and benefits. British Medical Journal 331:1169, 2005

Glassman AH, Bigger JT, Jr.: Antipsychotic drugs: prolonged QTc interval, torsade de pointes, and sudden death. American Journal of Psychiatry 158:1774-1782, 2001

Goff DC: New insights into clinical response in schizophrenia: from dopamine D2 receptor occupancy to patients' quality of life. American Journal of Psychiatry 165:940-943, 2008

Goldberg JF, Perlis RH, et al.: Adjunctive antidepressant use and symptomatic recovery among bipolar depressed patients with concomitant manic symptoms: findings from the STEP-BD. American Journal of Psychiatry 164:1348-1355, 2007

Gonzales D, Rennard SI, et al.: Varenicline, an alpha4beta2 nicotinic acetylcholine receptor partial agonist, vs. sustained-release bupropion and placebo for smoking cessation: a randomized controlled trial. Journal of the American Medical Association 296:47-55, 2006

Grunder G, Fellows C, et al.: Brain and plasma pharmacokinetics of aripiprazole in patients with schizophrenia: an [18F]fallypride PET study. American Journal of Psychiatry 165:988-995, 2008

Grunebaum MF, Ellis SP, et al.: Antidepressants and suicide risk in the United States, 1985-1999. Journal of Clinical Psychiatry 65:1456-1462, 2004

Grunze H, Langosch J, et al.: Levetiracetam in the treatment of acute mania: an open add-on study with an on-off-on design. Journal of Clinical Psychiatry 64:781-784, 2003

Grunze HC, Normann C, et al.: Antimanic efficacy of topiramate in 11 patients in an open trial with an on-off-on design. Journal of Clinical Psychiatry 62:464-468, 2001

Hall RC, Popkin MK, et al.: Physical illness presenting as psychiatric disease. Archives of General Psychiatry 35:1315-1320, 1978

Hamoda HM, Osser DN: The psychopharmacology algorithm project at the Harvard South Shore program: an update on psychotic depression. Harvard Review of Psychiatry 16:235-247, 2008

Hanley MJ, Kenna GA: Quetiapine: treatment for substance abuse and drug of abuse. American Journal of Health-System Pharmacy 65:611-618, 2008

Harada T, Sakamoto K, et al.: Incidence and predictors of activation syndrome induced by antidepressants. Depression and Anxiety 25:1014-1019, 2008

Heres S, Davis J, et al.: Why olanzapine beats risperidone, risperidone beats quetiapine, and quetiapine beats olanzapine: an exploratory analysis of head-to-head comparison studies of second-generation antipsychotics. American Journal of Psychiatry 163:185-194, 2006

Herrmann N, Lanctot KL: Do atypical antipsychotics cause stroke? CNS Drugs 19:91-103, 2005

Higgins ES: A comparative analysis of antidepressants and stimulants for the treatment of adults with attention-deficit hyperactivity disorder. Journal of Family Practice 48:15-20, 1999

Hu SC, Frucht SJ: Emergency treatment of movement disorders. Current Treatment Options in Neurology 9:103-114, 2007

Hughes JC, Cook CC: The efficacy of disulfiram: a review of outcome studies. Addiction 92: 381-395, 1997

Hyman SE, Arana GW, et al.: Handbook of Psychiatric Drug Therapy, Third Edition. Boston: Little, Brown and Company, 1995

IMS Health: 2007 Top therapeutic classes by U.S. dispensed prescriptions. www.imshealth.com. Retrieved October 12, 2008

Janicak PG, Davis JM, et al.: Principles and Practice of Psychopharmacotherapy. Philadelphia, PA: Lippincott Williams & Wilkins, 2006

Jefferson JW, Greist AH, et al.: Lithium Encyclopedia for Clinical Practice. Washington, D.C.: American Psychiatric Association Press, 1987

Joffe H, Cohen LS, et al.: Valproate is associated with new-onset oligoamenorrhea with hyperandrogenism in women with bipolar disorder. Biological Psychiatry 59:1078-1086, 2006

Johannessen CU: Mechanisms of action of valproate: a commentary. Neurochemistry International 37:103-110, 2000

Johannessen Landmark C: Antiepileptic drugs in non-epilepsy disorders: relations between mechanism of action and clinical efficacy. CNS Drugs 22:27-47, 2008

Johnson BA, Ait-Daoud N, et al.: Oral topiramate for treatment of alcohol dependence: a randomized controlled trial. Lancet 361:1677-1685, 2003

Johnson BA, Ait-Daoud N, et al.: Combining ondansetron and naltrexone effectively treats biologically predisposed alcoholics: from hypothesis to preliminary clinical evidence. Alcoholism: Clinical and Experimental Research 24:737-742, 2000

Johnson BA, Roache JD, et al.: Ondansetron for reduction of drinking among biologically predisposed alcoholic patients: A randomized controlled trial. Journal of the American Medical Association 284:963-971, 2000

Johnson BA, Rosenthal N, et al.: Topiramate for treating alcohol dependence: a randomized controlled trial. Journal of the American Medical Association 298:1641-1651, 2007

Johnson EM, Whyte E, et al.: Cardiovascular changes associated with venlafaxine in the treatment of late-life depression. American Journal of Geriatric Psychiatry 14:796-802, 2006

Johnson MW, Suess PE, et al.: Ramelteon: a novel hypnotic lacking abuse liability and sedative adverse effects. Archives of General Psychiatry 63:1149-1157, 2006

Johnston AM, Eagles JM: Lithium-associated clinical hypothyroidism. Prevalence and risk factors. British Journal of Psychiatry 175:336-339, 1999

Jorenby DE, Hays JT, et al.: Efficacy of varenicline, an alpha4beta2 nicotinic acetylcholine receptor partial agonist, vs. placebo or sustained-release bupropion for smoking cessation: a randomized controlled trial. Journal of the American Medical Association 296:56-63, 2006

Jorenby DE, Leischow SJ, et al.: A controlled trial of sustained-release bupropion, a nicotine patch, or both for smoking cessation. New England Journal of Medicine 340:685-691, 1999

Kane JM, Carson WH, et al.: Efficacy and safety of aripiprazole and haloperidol versus placebo in patents with schizophrenia and schizoaffective disorder. Journal of Clinical Psychiatry 63:763-771, 2002

Kane JM, Meltzer HY, et al.: Aripiprazole for treatment-resistant schizophrenia: results of a multicenter, randomized, double-blind, comparison study versus perphenazine. Journal of Clinical Psychiatry 68:213-223, 2007

Katon W, Von Korff M, et al.: Stepped collaborative care for primary care patients with persistent symptoms of depression: a randomized trial. Archives of General Psychiatry 56:1109-1115, 1999

Keck PE, Jr., Strawn JR, et al.: Pharmacologic treatment considerations in co-occurring bipolar and anxiety disorders. Journal of Clinical Psychiatry 67(Suppl 1):8-15, 2006

Keefe RS, Bilder RM, et al.: Neurocognitive effects of antipsychotic medications in patients with chronic schizophrenia in the CATIE Trial. Archives of General Psychiatry 64:633-647, 2007

Keller MB, McCullough JP, et al.: A comparison of nefazodone, the cognitive behavioral-analysis system of psychotherapy, and their combination for the treatment of chronic depression. New England Journal of Medicine 342:1462-1470, 2000

Kim H, Lim SW, et al.: Monoamine transporter gene polymorphisms and antidepressant response in Koreans with late-life depression. Journal of the American Medical Association 296:1609-1618, 2006

King M, Sibbald B, et al.: Randomized controlled trial of non-directive counseling, cognitive-behavior therapy and usual general practitioner care in the management of depression as well as mixed anxiety and depression in primary care. Health Technology Assessment 4:1-83, 2000

Kinon BJ, Volavka J, et al.: Standard and higher dose of olanzapine in patients with schizophrenia or schizoaffective disorder: a randomized, double-blind, fixed-dose study. Journal of Clinical Psychopharmacology 28:392-400, 2008

Kintz P: Deaths involving buprenorphine: a compendium of French cases. Forensic Science International 121:65-69, 2001

Kirchmayer U, Davoli M, et al.: Naltrexone maintenance treatment for opioid dependence. Cochrane Database of Systematic Reviews CD001333, 2003

Kleindienst N, Severus WE, et al.: Are serum lithium levels related to the polarity of recurrence in bipolar disorders? Evidence from a multicenter trial. International Clinical Psychopharmacology 22:125-131, 2007

Kleindienst N, Severus WE, et al.: Is polarity of recurrence related to serum lithium level in patients with bipolar disorder? European Archives of Psychiatry and Clinical Neuroscience 255:72-74, 2005

Ko DT, Hebert PR, et al.: Beta-blocker therapy and symptoms of depression, fatigue, and sexual dysfunction. Journal of the American Medical Association 288:351-357, 2002

Kohen I, Kremen N: Varenicline-induced manic episode in a patient with bipolar disorder. American Journal of Psychiatry 164:1269-1270, 2007

Kraft JB, Peters EJ, et al.: Analysis of association between the serotonin transporter and antidepressant response in a large clinical sample. Biological Psychiatry 61:734-742, 2007

Kranzler HR, Van Kirk J: Efficacy of naltrexone and acamprosate for alcoholism treatment: a meta-analysis. Alcoholism: Clinical and Experimental Research 25:1335-1341, 2001

Kuhn R: The treatment of depressive states with G 22355 (imipramine hydrochloride). American Journal of Psychiatry 115:459-464, 1958

Laaksonen E, Koski-Jannes A, et al.: A randomized, multicentre, open-label, comparative trial of disulfiram, naltrexone and acamprosate in the treatment of alcohol dependence. Alcohol and Alcoholism 43:53-61, 2008

Lader M, Scotto JC: A multicenter double-blind comparison of hydroxyzine, buspirone and placebo in patients with generalized anxiety disorder. Psychopharmacology 139:402-406, 1998

Lamberti JS, Olson D, et al.: Prevalence of the metabolic syndrome among patients receiving clozapine. American Journal of Psychiatry 163:1273-1276, 2006

Lekman M, Paddock S, et al.: Pharmacogenetics of major depression: insights from level 1 of the Sequenced Treatment Alternatives to Relieve Depression (STAR*D) trial. Molecular Diagnosis and Therapy 12:321-330, 2008

Lenox RH, Hahn CG: Overview of the mechanism of action of lithium in the brain: fifty-year update. Journal of Clinical Psychiatry 61(Suppl 9):5-15, 2000

Lepkifker E, Sverdlik A, et al.: Renal insufficiency in long-term lithium treatment. Journal of Clinical Psychiatry 65:850-856, 2004

Leucht S, Corves C, et al.: Second-generation versus first-generation antipsychotic drugs for schizophrenia: a meta-analysis. Lancet 373:31-41, 2009

Leverich GS, Altshuler LL, et al.: Risk of switch in mood polarity to hypomania or mania in patients with bipolar depression during acute and continuation trials of venlafaxine, sertraline, and bupropion as adjuncts to mood stabilizers. American Journal of Psychiatry 163:232-239, 2006

Lewis SW, Barnes TR, et al.: Randomized controlled trial of effect of prescription of clozapine versus other second-generation antipsychotic drugs in resistant schizophrenia. Schizophrenia Bulletin 32:715-723, 2006

Liappas IA, Malitas PN, et al.: Zolpidem dependence case series: possible neurobiological mechanisms and clinical management. Journal of Psychopharmacology 17:131-135, 2003

Lieberman JA, Stroup TS, et al.: Effectiveness of antipsychotic drugs in patients with chronic schizophrenia. New England Journal of Med 353:1209-1223, 2005

Ling W, Wesson RD: Naltrexone treatment for addicted health-care professionals: a collaborative private practice experience. Journal of Clinical Psychiatry 45:46-48, 1984

Lingford-Hughes AR, Welch S, et al.: Evidence-based guidelines for the pharmacological management of substance misuse, addiction and comorbidity: recommendations from the British Association for Psychopharmacology. Journal of Psychopharmacology 18:293-335, 2004

Lipkovich I, Citrome L, et al.: Early predictors of substantial weight gain in bipolar patients treated with olanzapine. Journal of Clinical Psychopharmacology 26:316-320, 2006

Lippman SB, Nash K: Monoamine oxidase inhibitor update. Potential adverse food and drug interactions. Drug Safety 5:195-204, 1990

Littleton J, Zieglgansberger W: Pharmacological mechanisms of naltrexone and acamprosate in the prevention of relapse in alcohol dependence. American Journal on Addictions 12(Suppl 1):S3-S11, 2003

Llorca PM, Spadone C, et al.: Efficacy and safety of hydroxyzine in the treatment of generalized anxiety disorder: a 3-month double blind study. Journal of Clinical Psychiatry 63:1020-1027, 2002

Lonergan E, Britton AM, et al.: Antipsychotics for delirium. Cochrane Database of Systematic Reviews CD005594, 2007

Macdonald KJ, Young LT: Newer antiepileptic drugs in bipolar disorder: rationale for use and role in therapy. CNS Drugs 16:549-562, 2002

Magni G: The use of antidepressants in the treatment of chronic pain. A review of the current evidence. Drugs 42:730-748, 1991

Malhotra AK, Murphy GM, Jr. et al.: Pharmacogenetics of psychotropic drug response. American Journal of Psychiatry 161:780-796, 2004

Mamo D, Graff A, et al.: Differential effects of aripiprazole on D(2), 5-HT(2), and 5-HT(1A) receptor occupancy in patients with schizophrenia: a triple tracer PET study. American Journal of Psychiatry 164:1411-1417, 2007

Mann K, Lehert P, et al.: The efficacy of acamprosate in the maintenance of abstinence in alcohol-dependent individuals: results of a meta-analysis. Alcoholism: Clinical and Experimental Research 28:51-63, 2004

Marangell LB, Dennehy EB, et al.: Case-control analyses of the impact of pharmacotherapy on prospectively observed suicide attempts and completed suicides in bipolar disorder: findings from STEP-BD. Journal of Clinical Psychiatry 69:916-922, 2008

Max MB, Culnane M, et al.: Amitriptyline relieves diabetic neuropathy pain in patients with normal or depressed mood. Neurology 37:589-596, 1987

Mbaya P, Alam F, et al.: Cardiovascular effects of high dose venlafaxine XL in patients with major depressive disorder. Human Psychopharmacology 22:129-133, 2007

McCue RE, Waheed R, et al.: Comparative effectiveness of second-generation antipsychotics and haloperidol in acute schizophrenia. British Journal of Psychiatry 189:433-440, 2006

McEvoy JP, Stiller RL, et al.: Plasma haloperidol levels drawn at neuroleptic threshold doses: a pilot study. Journal of Clinical Psychopharmacology 6:133-138, 1986

McGrath PJ, Khan AY, et al.: Response to a selective serotonin reuptake inhibitor (citalopram) in major depressive disorder with melancholic features: A STAR*D report. Journal of Clinical Psychiatry 69:1847-1855, 2008

McGrath PJ, Stewart JW, et al.: Tranylcypromine versus venlafaxine plus mirtazapine following three failed antidepressant medication trials for depression: a STAR*D report. American Journal of Psychiatry 163:1531-1541, 2006

McMahon FJ, Buervenich S, et al.: Variation in the gene encoding the serotonin 2A receptor is associated with outcome of antidepressant treatment. American Journal of Human Genetics 78:804-814, 2006

Megarbane B, Hreiche R, et al.: Does high-dose buprenorphine cause respiratory depression?: possible mechanisms and therapeutic consequences. Toxicological Reviews 25:79-85, 2006

Meltzer HY, Alphs L, et. al.: Clozapine treatment for suicidality in schizophrenia: International Suicide

Prevention Trial (InterSePT). Archives of General Psychiatry 60:82-91, 2003

Meszaros A, Czobor P, et al.: Pharmacotherapy of adult Attention Deficit/Hyperactivity Disorder (ADHD): a systematic review. Psychiatria Hungarica 22:259-270, 2007

Meyer JM, Simpson GM: From chlorpromazine to olanzapine: a brief history of antipsychotics. Psychiatric Services 48:1137-1139, 1997

Miceli JJ, Glue P, et al.: The effect of food on the absorption of oral ziprasidone. Psychopharmacology Bulletin 40:58-68, 2007

Michelson D, Adler L, et al.: Atomoxetine in adults with ADHD: two randomized, placebo-controlled studies. Biological Psychiatry 53:112-120, 2003

Miklowitz DJ: Adjunctive psychotherapy for bipolar disorder: state of the evidence. American Journal of Psychiatry 165:1408-1419, 2008

Miklowitz DJ, Otto MW, et al.: Psychosocial treatments for bipolar depression: a 1-year randomized trial from the Systematic Treatment Enhancement Program. Archives of General Psychiatry 64:419-426, 2007

Miller LJ: Prazosin for the treatment of posttraumatic stress disorder sleep disturbances. Pharmacotherapy 28:656-666, 2008

Montejo AL, Llorca G, et al.: Incidence of sexual dysfunction associated with antidepressant agents: a prospective multicenter study of 1022 outpatients. Spanish Working Group for the Study of Psychotropic-Related Sexual Dysfunction. Journal of Clinical Psychiatry 62(Suppl 3):10-21, 2001

Morley KC, Teesson M, et al.: Naltrexone versus acamprosate in the treatment of alcohol dependence: A multi-centre, randomized, double-blind, placebo-controlled trial. Addiction 101:1451-1462, 2006

Morrato EH, Libby AM, et al.: Frequency of provider contact after FDA advisory on risk of pediatric suicidality with SSRIs. American Journal of Psychiatry 165:42-50, 2008

Murphy GM, Jr., Hollander SB, et al.: Effects of the serotonin transporter gene promoter polymorphism on mirtazapine and paroxetine efficacy and adverse events in geriatric major depression. Archives of General Psychiatry 61:1163-1169, 2004

Nestler EJ, Hyman SE, Malenka RC: Molecular Neuropharmacology: A Foundation for Clinical Neuroscience, Second Edition. New York: McGraw-Hill Companies, Inc., 2009

Nierenberg AA, Fava M, et al.: A comparison of lithium and T(3) augmentation following two failed medication treatments for depression: a STAR*D report. American Journal of Psychiatry 163:1519-1530, 2006

Nunes PV, Forlenza OV, et al.: Lithium and risk for Alzheimer's disease in elderly patients with bipolar disorder. British Journal of Psychiatry 190:359-360, 2007

Nutt DJ, Malizia AL: New insights into the role of the GABA(A)-benzodiazepine receptor in psychiatric disorder. British Journal of Psychiatry 179:390-396, 2001

Nyberg S, Eriksson B, et al.: Suggested minimal effective dose of risperidone based on PET-measured D2 and 5-HT2A receptor occupancy in schizophrenic patients. American Journal of Psychiatry 156:869-875, 1999

O'Donovan C, Garnham JS, et al.: Antidepressant monotherapy in pre-bipolar depression; predictive value and inherent risk. Journal of Affective Disorders 107:293-298, 2008

O'Donovan C, Kusumakar V, et al.: Menstrual abnormalities and polycystic ovary syndrome in women taking valproate for bipolar mood disorder. Journal of Clinical Psychiatry 63:322-330, 2002

O'Malley SS, Garbutt JC, et al.: Efficacy of extended-release naltrexone in alcohol-dependent patients who are abstinent before treatment. Journal of Clinical Psychopharmacology 27:507-512, 2007

Olfson M, Marcus SC, et al.: Treatment of schizophrenia with long-acting fluphenazine, haloperidol, or risperidone. Schizophrenia Bulletin 33:1379-1387, 2007

Onghena P, Van Houdenhove B: Antidepressant-induced analgesia in chronic non-malignant pain: a meta-analysis of 39 placebo-controlled studies. Pain 49:205-219, 1992

Ortenzi A, Paggi A, et al.: Oxcarbazepine and adverse events: impact of age, dosage, metabolite serum concentrations and concomitant antiepileptic therapy. Functional Neurology 23:97-100, 2008

Osser DN: Cleaning up evidence-based psychopharmacology. Psychopharm Review 43:19-25, 2008

Osser DN, Najarian DM, Dufresne RL: Olanzapine increases weight and serum triglyceride levels. Journal of Clinical Psychiatry 60:767-770, 1999

Osser DN, Sigadel R: Short-term inpatient pharmacotherapy of schizophrenia. Harvard Review of Psychiatry 9:89-104, 2001

Pande AC, Crockatt JG, et al.: Gabapentin in bipolar disorder: a placebo-controlled trial of adjunctive therapy. Gabapentin Bipolar Disorder Study Group. Bipolar Disorders 2:249-255, 2000

PDR-Physicians Desk Reference Concise Prescribing Guide. Montvale, NJ: Thomson Reuters, Issue 3, 2008

Perry PJ, Zeilmann C, et al.: Tricyclic antidepressant concentrations in plasma: an estimate of their sensitivity and specificity as a predictor of response. Journal of Clinical Psychopharmacology 14:230-240, 1994

Perucca E: Clinically relevant drug interactions with antiepileptic drugs. British Journal of Clinical Pharmacology 61:246-255, 2006

Pettinati HM, O'Brien CP, et al.: The status of naltrexone in the treatment of alcohol dependence: specific effects on heavy drinking. Journal of Clinical Psychopharmacology 26:610-625, 2006

Phansalkar S, Osser DN: Optimizing Clozapine Treatment: Part I. Psychopharm Review 44:1-8, 2009

Phansalkar S, Osser DN: Optimizing Clozapine Treatment: Part II. Psychopharm Review 44:9-15, 2009

Phelps J: The bipolar spectrum, in Parker G (ed.), Bipolar II Disorder. Modeling, Measuring, and Managing. Cambridge, UK: Cambridge University Press, 2008

Pigott TA, Carson WH, et al.: Aripiprazole for the prevention of relapse in stabilized patients with chronic schizophrenia: a placebo-controlled 26-week study. Journal of Clinical Psychiatry 64:1048-1056, 2003

Popper CW: Antidepressants in the treatment of attention-deficit/hyperactivity disorder. Journal of Clinical Psychiatry 58(Suppl 14):14-29, 1997

Post RM, Altshuler LL, et al.: Mood switch in bipolar depression: comparison of adjunctive venlafaxine, bupropion and sertraline. British Journal of Psychiatry 189:124-131, 2006

Pratoomsri W, Yatham LN, et al.: Oxcarbazepine in the treatment of bipolar disorder: a review. Canadian Journal of Psychiatry 51:540-545, 2006

Prince JB, Wilens TE, et al.: A controlled study of nortriptyline in children and adolescents with attention deficit hyperactivity disorder. Journal of Child and Adolescent Psychopharmacology 10:193-204, 2000

Quitkin FM, Stewart JW, et al.: Columbia atypical depression. A subgroup of depressives with better response to MAOI than to tricyclic antidepressants or placebo. British Journal of Psychiatry. (Suppl 21):30-34, 1993

Raja M: Improvement or worsening of psychotic symptoms after treatment with low doses of aripiprazole. International Journal of Neuropsychopharmacology 10:107-110, 2007

Raskin J, Goldstein DJ, et al.: Duloxetine in the long-term treatment of major depressive disorder. Journal of Clinical Psychiatry 64:1237-1244, 2003

Raskind MA, Peskind ER, et al.: A parallel group placebo controlled study of prazosin for trauma nightmares and sleep disturbance in combat veterans with post-traumatic stress disorder. Biological Psychiatry 61:928-934, 2007

Ray WA, Chung CP, et al.: Atypical antipsychotic drugs and the risk of sudden cardiac death. New England Journal of Medicine 360:225-235, 2009

Ray WA, Meredith S, et al.: Cyclic antidepressants and the risk of sudden cardiac death. Clinical Pharmacology and Therapeutics 75:234-241, 2004

Reynolds CF, Frank E, et al.: Nortriptyline and interpersonal psychotherapy as maintenance therapies for recurrent major depression: a randomized controlled trial in patients older than 59 years. Journal of the American Medical Association 281:39-45, 1999

Richelson E: Interactions of antidepressants with neurotransmitter transporters and receptors and their clinical relevance. Journal of Clinical Psychiatry 64(Suppl 13):5-12, 2003

Rochon PA, Normand SL, et al.: Antipsychotic therapy and short-term serious events in older adults with dementia. Archives of Internal Medicine 168:1090-1096, 2008

Rosenbaum JF, Arana GW, et al.: Handbook of Psychiatric Drug Therapy, Fifth Edition. Philadelphia, PA: Lippincott Williams & Wilkins, 2005

Roth T, Seiden D, et al.: Effects of Ramelteon on patient-reported sleep latency in older adults with chronic insomnia. Sleep Medicine 7:312-318, 2006

Rubio G, Jimenez-Arriero MA, et al.: Naltrexone versus acamprosate: one year follow-up of alcohol dependence treatment. Alcohol and Alcoholism 36:419-425, 2001

Rush AJ, Trivedi MH, et al.: Acute and longer-term outcomes in depressed outpatients requiring one or several treatment steps: a STAR*D report. American Journal of Psychiatry 163:1905-1917, 2006

Saarto T, Wiffen PJ: Antidepressants for neuropathic pain. Cochrane Database of Systematic Reviews CD005454, 2007

Sachs GS, Nierenberg AA, et al.: Effectiveness of adjunctive antidepressant treatment for bipolar depression. New England Journal of Medicine 356:1711-1722, 2007

Sanger DJ: The pharmacology and mechanisms of action of new generation, non-benzodiazepine hypnotic agents. CNS Drugs 18(Suppl 1):9-15, 2004

Sass H, Soyka M, et al.: Relapse prevention by acamprosate. Results from a placebo-controlled study on alcohol dependence. Archives of General Psychiatry 53:673-680, 1996

Sateia MJ, Kirby-Long P, et al.: Efficacy and clinical safety of Ramelteon: an evidence-based review. Sleep Medicine Reviews 12:319-332, 2008

Satel SL, Nelson JC: Stimulants in the treatment of depression: a critical overview. Journal of Clinical Psychiatry 50:241-249, 1989

Satterthwaite TD, Wolf DH, et al.: A meta-analysis of the risk of acute extrapyramidal symptoms with intramuscular antipsychotics for the treatment of agitation. Journal of Clinical Psychiatry 69:1869-1879, 2008

Schneeweiss S, Setoguchi S, et al.: Risk of death associated with the use of conventional versus atypical antipsychotic drugs among elderly patients. Canadian Medical Association Journal 176:627-632, 2007

Schneider LS, Dagerman KS, et al.: Risk of death with atypical antipsychotic drug treatment for dementia: meta-analysis of randomized placebo-controlled trials. Journal of the American Medical Association 294:1934-1943, 2005

Schneider LS, Tariot PN, et al.: Effectiveness of atypical antipsychotic drugs in patients with Alzheimer's disease. New England Journal of Medicine 355:1525-1538, 2006

Seeman P: Atypical antipsychotics: mechanism of action. Canadian Journal of Psychiatry 47:27-38, 2002

Sethi PK, Khandelwal DC: Zolpidem at supratherapeutic doses can cause drug abuse, dependence and withdrawal seizure. Journal of the Association of the Physicians of India 53:139-140, 2005

Severus WE, Kleindienst N, et al.: What is the optimal serum lithium level in the long-term treatment of bipolar disorder--a review? Bipolar Disorders 10:231-237, 2008

Shah RR: Drug-induced QT dispersion: does it predict the risk of torsade de pointes? Journal of Electrocardiology 38:10-18, 2005

Shaldubina A, Agam G, et al.: The mechanism of lithium action: state of the art, ten years later. Progress in Neuropsychopharmacology and Biological Psychiatry 25:855-866, 2001

Sikich L, Frazier JA, et al.: Double-blind comparison of first- and second-generation antipsychotics in early-onset schizophrenia and schizo-affective disorder: findings from the treatment of early-onset schizophrenia spectrum disorders (TEOSS) study. American Journal of Psychiatry 165:1420-1431, 2008

Silagy C, Lancaster T, et al.: Nicotine replacement therapy for smoking cessation. Cochrane Database of Systematic Reviews CD000146, 2004

Silagy C, Mant D, et al.: Nicotine replacement therapy for smoking cessation. Cochrane Database of Systematic Reviews CD000146, 2000

Siriwardena AN, Qureshi Z, et al.: GPs' attitudes to benzodiazepines and 'Z-drug' prescribing: a barrier to implementation of evidence and guidance on hypnotics. British Journal of General Practice 56:964-967, 2006

Sivertsen B, Omvik S, et al.: Cognitive behavioral therapy vs. zopiclone for treatment of chronic primary insomnia in older adults: a randomized controlled trial.

Journal of the American Medical Association 295:2851-2858, 2006

Soares-Weiser K, Fernandez HH: Tardive dyskinesia. Seminars in Neurology 27:159-169, 2007

Spina E, Scordo MG, et al.: Metabolic drug interactions with new psychotropic agents. Fundamental and Clinical Pharmacology 17:517-538, 2003

Srisurapanont M, Jarusuraisin N: Opioid antagonists for alcohol dependence. Cochrane Database of Systematic Reviews CD001867, 2005

Stagnitti MN: Antidepressants prescribed by medical doctors in office based and outpatient settings by specialty for the U.S. civilian non-institutionalized population, 2002 and 2005. Statistical Brief #206. Medical Expenditure Panel Survey. Agency for Healthcare Research and Quality. 2008

Stahl SM: Essential Psychopharmacology: The Prescriber's Guide. Cambridge, UK: Cambridge University Press, 2005

Stahl SM: Stahl's Essential Psychopharmacology: Neuroscientific Basis and Practical Applications. 3rd Edition. New York, NY: Cambridge University Press, 2008

Stahl SM, Grady MM: Differences in mechanism of action between current and future antidepressants. Journal of Clinical Psychiatry 64(Suppl 13):13-17, 2003

Sternbach H: The serotonin syndrome. American Journal of Psychiatry 148:705-713, 1991

Straus SM, Bleumink GS, et al.: Antipsychotics and the risk of sudden cardiac death. Archives of Internal Medicine 164:1293-1297, 2004

Suh JJ, Pettinati HM, et al.: The status of disulfiram: a half of a century later. Journal of Clinical Psychopharmacology 26:290-302, 2006

Sultzer DL, Davis SM, et al.: Clinical symptom responses to atypical antipsychotic medications in Alzheimer's disease: phase 1 outcomes from the CATIE-AD effectiveness trial. American Journal of Psychiatry 165:844-854, 2008

Summerfelt WT, Meltzer HY: Efficacy vs. effectiveness in psychiatric research. Psychiatric Services 49:834-835, 1998

Suzuki T, Uchida H, et al.: How effective is it to sequentially switch among olanzapine, quetiapine and risperidone?--A randomized, open-label study of algorithm-based antipsychotic treatment to patients with symptomatic schizophrenia in the real-world clinical setting. Psychopharmacology 195:285-295, 2007

Taylor D, Paton C, Kerwin R: The South London and Maudsley NHS Foundation Trust Oxleas NHS Foundation Trust Prescribing Guidelines, 9th Edition. London: Informa Healthcare, Telephone House, 2007

Taylor FB, Lowe K, et al.: Daytime prazosin reduces psychological distress to trauma specific cues in civilian trauma posttraumatic stress disorder. Biological Psychiatry 59:577-581, 2006

Taylor FB, Martin P, et al.: Prazosin effects on objective sleep measures and clinical symptoms in civilian trauma posttraumatic stress disorder: a placebo-controlled study. Biological Psychiatry 63:629-632, 2008

Thase ME, Macfadden W, et al.: Efficacy of quetiapine monotherapy in bipolar I and II depression: a double-blind, placebo-controlled study (the BOLDER II study). Journal of Clinical Psychopharmacology 26:600-609, 2006

Tonstad S, Tonnesen P, et al.: Effect of maintenance therapy with varenicline on smoking cessation: a randomized controlled trial. Journal of the American Medical Association 296:64-71, 2006

Trivedi M, Thase ME, et al.: Adjunctive aripiprazole in major depressive disorder: analysis of efficacy and safety in patients with anxious and atypical features. Journal of Clinical Psychiatry 69:1928-1936, 2008

Trivedi MH, Fava M, et al.: Medication augmentation after the failure of SSRIs for depression. New England Journal of Medicine 354:1243-1252, 2006

Van Winkel R, De Hert M, et al.: Screening for diabetes and other metabolic abnormalities in patients with schizophrenia and schizoaffective disorder: evaluation of incidence and screening methods. Journal of Clinical Psychiatry 67:1493-1500, 2006

Victorri-Vigneau C, Dailly E, et al.: Evidence of zolpidem abuse and dependence: results of the French Centre for Evaluation and Information on Pharmacodependence (CEIP) network survey. British Journal of Clinical Pharmacology 64:198-209, 2007

Vieta E, Goikolea JM, et al.: 1-year follow-up of patients treated with risperidone and topiramate for a manic episode. Journal of Clinical Psychiatry 64:834-839, 2003

Vieta E, Manuel Goikolea, J, et al.: A double-blind, randomized, placebo-controlled, prophylaxis study of adjunctive gabapentin for bipolar disorder. Journal of Clinical Psychiatry 67:473-477, 2006

Vieta E, Sanchez-Moreno J, et al.: Adjunctive topiramate in bipolar II disorder. World Journal of Biological Psychiatry 4:172-176, 2003

Viguera AC, Koukopoulos A, et al.: Teratogenicity and anticonvulsants: lessons from neurology to psychiatry. Journal of Clinical Psychiatry 68(Suppl 9):29-33, 2007

Waal HJ: Propranolol-induced depression. British Medical Journal 2:50, 1967

Wagner AK, Zhang F, et al.: Benzodiazepine use and hip fractures in the elderly: who is at greatest risk? Archives of Internal Medicine 164:1567-1572, 2004

Wang PS, Schneeweiss S, et al.: Risk of death in elderly users of conventional vs. atypical antipsychotic medications. New England Journal of Medicine 353:2335-2341, 2005

Washton AM, Gold MS, et al.: Successful use of naltrexone in addicted physicians and business executives. Advances in Alcohol and Substance Abuse 4:89-96, 1984

Weisler RH, Kalali AH, et al.: A multicenter, randomized, double-blind, placebo-controlled trial of extended-release carbamazepine capsules as monotherapy for bipolar disorder patients with manic or mixed episodes. Journal of Clinical Psychiatry 65:478-484, 2004

Weisler RH, Keck PE, et al.: Extended-release carbamazepine capsules as monotherapy for acute mania in bipolar disorder: a multicenter, randomized, double-blind, placebo-controlled trial. Journal of Clinical Psychiatry 66:323-330, 2005

Whitworth AB, Fischer F, et al.: Comparison of acamprosate and placebo in long-term treatment of alcohol dependence. Lancet 347:1438-1442, 1996

WHO-World Health Organization: WHO Model List of Essential Medicines, 15th list. www.who.int/medicines/publications/essentialmedicines/en/index.html. Retrieved November 2008

Wilens TE: Impact of ADHD and its treatment on substance abuse in adults. Journal of Clinical Psychiatry 65(Suppl 3):38-45, 2004

Wilens TE, Biederman J, et al.: Six-week, double-blind, placebo-controlled study of desipramine for adult attention deficit hyperactivity disorder. American Journal of Psychiatry 153:1147-1153, 1996

Wilens TE, Faraone SV, et al.: Does stimulant therapy of attention-deficit/hyperactivity disorder beget later substance abuse? A meta-analytic review of the literature. Pediatrics 111:179-185, 2003

Wisniewski SR, Fava M, et al.: Acceptability of second-step treatments to depressed outpatients: a STAR*D report. American Journal of Psychiatry 164:753-760, 2007

Wohlreich MM, Mallinckrodt CH, et al.: Duloxetine for the treatment of major depressive disorder: safety and tolerability associated with dose escalation. Depression and Anxiety 24:41-52, 2007

Wu RR, Zhao JP, et al.: Lifestyle intervention and metformin for treatment of antipsychotic-induced weight gain: a randomized controlled trial. Journal of the American Medical Association 299:185-193, 2008

Yatham LN, Kusumakar V, et al.: Third generation anticonvulsants in bipolar disorder: a review of efficacy and summary of clinical recommendations. Journal of Clinical Psychiatry 63:275-283, 2002

Young AH, Geddes JR, et al.: Tiagabine in the maintenance treatment of bipolar disorders. Cochrane Database of Systematic Reviews CD005173, 2006

Young AH, Geddes JR, et al.: Tiagabine in the treatment of acute affective episodes in bipolar disorder: efficacy and acceptability. Cochrane Database of Systematic Reviews CD004694, 2006

Yury CA, Fisher JE: Meta-analysis of the effectiveness of atypical antipsychotics for the treatment of behavioral problems in persons with dementia. Psychotherapy and Psychosomatics 76:213-218, 2007

Zacher JL, Roche-Desilets J: Hypotension secondary to the combination of intramuscular olanzapine and intramuscular lorazepam. Journal of Clinical Psychiatry 66:1614-1615, 2005

Made in the USA
Lexington, KY
27 March 2013